OUTER SPACE
LAW, POLICY AND GOVERNANCE

OUTER SPACE
LAW, POLICY AND GOVERNANCE

G.S. Sachdeva M.A., LL.B., Ph.D.

Introduction by
Air Marshal **Vinod Patney** SYSM PVSM AVSM VrC (Retd)

KW Publishers Pvt Ltd
New Delhi

in association with

Centre for Air Power Studies
New Delhi

2011 BEST PUBLISHERS AWARD (ENGLISH)

First reprint March 2014
Second reprint June 2016

The Centre for Air Power Studies is an independent, non-profit, academic research institution established in 2002 under a registered Trust to undertake and promote policy-related research, study and discussion on the trends and developments in defence and military issues, especially air power and the aerospace arena, for civil and military purposes. Its publications seek to expand and deepen the understanding of defence, military power, air power and aerospace issues without necessarily reflecting the views of any institution or individuals except those of the authors.

Centre for Air Power Studies
P-284, Arjan Path
Subroto Park
New Delhi 110010

Tele: (91-11) 25699131
E-mail: diroffice@aerospaceindia.org

© 2014, G.S. Sachdeva

All rights reserved. No part of this book may be reproduced or transmitted in any form or by any means, electronic or mechanical, including photocopying, recording or by any information storage and retrieval system, without permission in writing.

Published in India by
Kalpana Shukla
KW Publishers Pvt Ltd
4676/21, First Floor, Ansari Road, Daryaganj, New Delhi 110002
T: +91 11 23263498 / 43528107 E: knowledgeworld@kwpub.com • www.kwpub.com

ISBN 978-93-81904-88-6

Printed by Bhavish Graphics, Chennai

Contents

	Preface	ix
	Introduction	xiii
1.	*Jus Cogens* of Space Law: A Proposal	1
2.	Astronauts as Envoys of Mankind in Outer Space: Resolution of a Dilemma	31
3.	Space Policy of India: A Few Pointers	63
4.	India as Vendor of Space Utilities to Developing Countries: An Example in Cooperation	101
5.	International Cooperation as Core Concept of Space Law: For Diplomacy and Confidence	123
6.	Mining of Asteroids: A Legal Analysis for Effective Governance	155
7.	International Code of Conduct for Activities in Outer Space: An Exercise in Futility	205

Annexure A
Treaty on Principles Governing the Activities of States in the Exploration and Use of Outer Space, including the Moon and other Celestial Bodies — 233

Annexure B
Agreement Governing the Activities of States on the Moon and Other Celestial Bodies (1979) — 241

Index — 253

DEDICATION

To my Parents
Sardar Sapuran Singh and Bibi Vidwant Kaur

And my Uncle
Lt Col Khazan Singh Sachdeva (Retd)

Preface

Space law evolved out of International Law as a necessity when man started sending artificial objects into outer space. The purpose of Space law was to maintain public order in outer space and ensure good governance of the new domain. Initially the international comity of states, buoyed by the euphoria of so called conquest of outer space, started negotiating innovative legal concepts in a display of great amity and rare unanimity towards common objective. But this gusto petered out within a decade. No formal treaty or convention on space issues has been concluded since the Moon Agreement of 1979. No wonder, some gaps in space law were deliberately left due to lack of consensus since the major space powers preferred hedging strategies over closing options, some inadvertently crept in due to lack of vision, and some other less understood concepts were pended till better understanding of space phenomena or development of supportive technology. These have since started jumping to centre-stage and constitute the contemporary challenges. With all humility, this book seeks to grapple with this niche of deficiencies and hopes to excite a scholarly debate on nascent issues.

Developments in space technology have been fast and spectacular over the last few years. Space tourism which seemed a dream just two decades back has turned a reality and may become affordable to the middle class in another decade. Scientific surveys and robotic studies on asteroids have revealed massive reserves of natural resources in mineral and metal. Their economic exploitation with viable technologies seemed an almost insurmountable task barely a few years back. Today, it looks promising and profitable. Remote sensing has gone many a miles with clear imagery and inexpensively sensed data. Satellite communication and navigation have been working in our service for decades; and the list can be continued. This techno-scientific reality overwhelms and shocks us to realize that law is a laggard, even on its own turf, and never proactive enough to prance hand in hand with technology.

This book is a wake up call on some of these challenges and unresolved contingencies that failed consensus during negotiations but today stare us in the face to be addressed suitably. For example, ambiguity on policy issues like international cooperation in outer space and futility of reiterative codes of conduct as soft law adjuncts, absence of legal regimen governing mining on the asteroids by private enterprise in competitive environment, explication of the concept of Common Heritage of Mankind and modalities for sharing of benefits between nations, procedures for nomination of an astronaut as envoy of mankind in space by name, designation or seniority to avoid confusion in eventualities of multiple astronauts of multiple nationalities, all being at one place at the same time, *et al.*

The agenda for this book is modest. It does not envisage a comprehensive coverage of chinks in space law. Rather, it comprises a limited number of topics like space jurisprudence, applicability of international law to space sphere, space policy *vis a vis* soft law codes and issues of effective governance in outer space. The book has attempted to discuss mentioned points and has come to the conclusion that effective public order in outer space, good governance of celestial resources and regulation of space activities through law can lead to progressively better welfare and perpetual growth in prosperity of humanity. However, some of the issues analysed may appear controversial or solutions offered seemingly unviable. In such cases, debate would be desirable to generate greater research and analysis by scholars to mull over the issues and arrive at judicious consensus. The author, in due humility, acknowledges his failings in research content, fallacies in analysis and frailties in the style of presentation.

Some months back, the manuscript of the book was accepted for publication by Air Cmde Jasjit Singh, Director General, Centre for Air Power Studies (CAPS), but it is a matter of great regret that he will not be there to release the book. The loss is further accentuated by his eternal absence and the book not having the privilege to bear his customary Introduction. My pain would be understandable in having lost a good guide and excellent friend. At the same time, India has been bereft of a gallant airman, seasoned strategist and an innovative thinker on defence matters. CAPS has lost its father-figure, a leader who built and nurtured this institute to its present stature of scholarship and reputation.

I too owe a substantial intellectual debt to Air Cmde Jasjit Singh in creation of this volume, and particularly for his constant guidance and prodding encouragement to complete this work. I must also acknowledge the invaluable help of Mr. Y. P. Madan and Dr. Manpreet Sethi in my research to hone finer details. This has substantially improved the quality of content and depth of knowledge on certain issues.

Profuse thanks are genuinely due to Ms. Kalpana Shukla and Mr. Jose Mathew of KW Publishers Pvt Ltd for their sincere efforts in pushing the deadlines for publication of this book. Ms Rehana Misra has done a commendable job of editing and I am indeed grateful to her for the pains she takes over details.

August 15, 2013   Dr. **G.S. Sachdeva**

Introduction

Vinod Patney

The English novelist George Orwell once remarked, "Who controls the past, controls the future. Who controls the present, controls the past." A valid statement. Power has long been recognised as the main arbiter of international relations. Arm-twisting, in its myriad forms, is the preferred method of exercising power. Two other truisms are worthy of mention. The first is that laws and policies adopted or accepted are indicative of the power equations of the time. Second, such laws can never be permanent but altering or amending them is, once again, a function of extant power equations. The nature and content of outer space treaties, and the fact that no amendments have been carried out in spite of so many intervening years is a reflection of the same above mentioned truisms.

The exploitation of space and its importance were long recognised. The unspoken race between the USA and the Soviet Union was much in evidence. The USSR took the first tiny step with the launch of the Sputnik satellite on October 4, 1957. Compared to the current utilisation, exploitation and future plans for space, it was indeed a 'baby' step. The Sputnik satellite was a polished metal sphere of a mere 58 cm diameter with four external radio antennae, and it was launched into a low earth orbit. The signals from the satellite lasted only till October 26, 1957, for only 22 days, as the batteries got discharged. Hence, the technological claim to fame of the Sputnik satellite was that it was the first satellite launched into space and not really for any other reason.

Air Marshal **Vinod Patney** SYSM PVSM AVSM VrC (Retd) is one of the most distinguished fighter pilots in the Indian Air Force's history: a decorated veteran of the 1965 and 1971 Wars. He was awarded the Sarvottam Yudh Seva Medal (SYSM) for spearheading air attacks during the Kargil War, making him the sole recipient of the SYSM till date in the IAF. Currently, he is the Director, Centre for Air Power Studies, New Delhi.

The impact of the Sputnik was remarkable, far in excess of the technological breakthrough. Space has always been recognised as the ultimate proverbial 'high ground'. At that time, the Cold War was unrelenting and good reconnaissance capabilities were at a premium. The USA accelerated its programme for space conquest and on January 31, 1958, the 18-pound satellite Explorer-1 was launched. Far more importantly, within a few weeks of the launch of the Sputnik, the USA formed the National Reconnaissance Organisation, an indication of the importance attached to space exploitation.

The impetus for space exploration picked up speed thereafter with the launch of a number of improved satellites but the next big milestone was the first manned space flight by Yuri Gagarin of the USSR on April 12, 1961. The USA soon launched its own manned space flights with a sub-orbital flight by Alan Shepard on May 5, 1961, and the first full orbital flight by John Glenn on February 20, 1962. The battle for exploitation of space was well and truly joined.

With the remarkable impetus to manned space flights given by President John F. Kennedy, the US was the first country to land a man, Neil Armstrong, on the Moon on July 20, 1969. The romance of space has not looked back since. Incidentally, even before the first Moon landing, in October 1963, a treaty banning nuclear weapon tests in the atmosphere in outer space and under water, was signed. More importantly, while the whole world was still marvelling at the human advent into outer space and the often vicarious sense of achievement, it dawned on strategic thinkers and law scholars that there was a need to regulate human activities in outer space to ensure public order. The need was also felt for a specialised law to govern outer space. Under the aegis of the United Nations General Assembly and its relevant committees, sterling contributions were made in this task. As a result, unique and innovative concepts were embodied in this corpus and space law was born

The Outer Space Treaty came into force in 1967 followed in succession by the other four Outer Space Treaties by 1979. A close examination of these treaties clearly reflects the politics of the time. Perforce, certain gaps remain, or have deliberately been retained, in the law. For instance, there is a lack of consensus on controversial issues and a lack of techno-vision on the future of space technology. The desired cooperation and motivation for political and diplomatic niceties were often not present, possibly because the negotiations were being held in the backdrop of the Cold War.

Over the next two decades or so, other institutions have also been created like the International Telecommunication Satellite Organisation and the Organisation for Space Communications.

Meanwhile, the progress in space exploration and exploitation has been remarkable. Space assets now have an overpowering influence on both civilian and military applications. A privately funded manned space flight was launched on June 20, 2004, and private enterprises are offering space flights and even one-way rides to Mars. Earth-launched spacecraft have now gone past the Earth's solar system to reconnoitre really distant areas. Militarisation of space took place many decades ago and weaponisation of space is under active discussion and research.

The body of space law that developed in the 1960s and 1970s, and which consists primarily of five international instruments is, however, beginning to reveal chinks in the face of rapid advances in space technology and the emerging realities of increasing government and non-government presence in outer space. Undoubtedly, treaties signed some five decades ago require considerable revision but efforts in this regard have not been encouraging.

Indeed, outer space is becoming more and more relevant to humanity and human welfare. Man is rapidly depleting the mineral resources on the Earth. The threat of looming scarcity is showing incipient signs of a scramble into outer space for mining for minerals and even setting up of infrastructure for manufacture of goods in a low gravity, clean environment. In view of this economic scenario, it can be visualised that the future star wars of the 21st century will be less for military objectives and more to capture economic resources. After all, nations have often gone to war to secure economic resources. Better governance through international law that ensures global sharing of benefits is the need of the hour.

In this backdrop, this book by Dr. G. S. Sachdeva, an expert on air and space law of longstanding scholarship, is most timely and relevant. His earlier book on *Outer Space Security and Legal Challenges* was well received. This book should equally excite informed readership.

Western and even Soviet strategists have for long articulated their national interests in outer space. We, in India, have been remiss on this count. Therefore, a book such as this that captures the evolution of the Space Policy in India, evaluates India's space activities over the last half a century and gropes for its direction, is of immense value. The author offers good hints for a future space strategy for India, including the military domain. He also espouses the case of India as a vendor of space utilities and space products to the developing nations, including the South Asian Association for Regional Cooperation (SAARC) fraternity, on a friendly and competitive basis. It vouchsafes India's pledge to cooperation in the space arena while safeguarding its national interests for sustainable and responsible utilisation of this global common.

This book is a good medley of essays on contemporary space concerns and makes for an astute analysis of the issues involved to arrive at constructive and viable solutions of durable value. It certainly furthers the objectives of the Centre for Air Power Studies to promote research in, and understanding of, strategic studies to assist policy decision-making.

I have little hesitation in recommending this book for a wide national and international readership of strategists, scholars and laymen who have interest in a subject whose importance can only increase with time.

Vinod Patney
Director, Centre for Air Power Studies

1

Jus Cogens of Space Law: A Proposal*

Introduction

Space law is a nascent and as yet an evolving branch of international law. Though developed as an adjunct of international law, it has traversed a journey of centuries in just decades to ripen. Its principles have matured into customs with obligatory force to elicit voluntary abidance. It has, thus, metabolised at an accelerated rate and is now getting metamorphosed into an independent and auto-poietic system[1] with a close nexus with, and cross-linkages to, other sub-systems of cognate legal regimes,[2] international jurisprudence, their intertwined operations and other multiple applications.

The evolutionary process undergone by international law finds a certain degree of parallel with that of the regime of space law *albeit* compressed in timeframe. In just half a century, the corpus of space law has accumulated one treaty, two conventions, and two agreements, two UN Declarations and four Principles and Guidelines on important space issues.[3] It is no small achievement through the endeavours of the UN Committee on Peaceful Uses of Outer Space (COPUOS) and the UN General Assembly in such a short span and on such contentious matters, considering that negotiation of treaties and conventions is a long and laborious affair. The efforts of COPUOS are continuing and deserve to be highly commended and duly supported by all participants.

Space law is gradually becoming complicated and its intrinsic complexity becomes evident as its contours gradually crystallise into new formations and its grammar becomes more distinct and communicable. However, there is no reductionist approach to criminality in space law; the

---

* A shorter version of this chapter has appeared as article in *Asian Journal of Air & Space Law*, Vol. II, No.2, July-December 2012. With the kind permission of Eastern Book Company.

bottom-line is either common survival or collective annihilation. The stakes are clear and the choices really limited. The decision, however, is of our volition and depends upon our level of sagacity. Therefore, strict adherence to space law is mandated to the international comity irrespective of their being space-farers, space-users or space-watchers.

An important international instrument called the Outer Space Treaty[4] (OST) is considered the *grundnorm* of space law as it delineates the basic tenets for human activities and stipulates the norms of conduct for the states in outer space. Whereas international law attempts to regulate relations among a society of states, space law transforms governance into a legal order of a genuine state community.[5] The difference may appear subtle yet the impact of this orientation is profound and positive. This treaty, thus, lays down the fundamental law for public order in outer space. By virtue of its universal mandate, fundamental contents and continued habitude, the OST has already gained the force of universality and assumed the status of customary law of outer space. In consequence, it is treated as legally binding even on states that are not party to the treaty or have not yet acceded to it.

Some of the salient provisions of the OST are so basic, fundamental and natural to the legal regime that these can be taken as peremptory norms of state behaviour and conduct in outer space and, hence, can logically be elevated to the pedestal of *jus cogens* of space law for universal obedience and impeccable compliance. A couple of examples would clarify the concept. First, outer space and the celestial bodies are free for exploration and use by all states on the basis of equality and activities here shall be carried out for the benefit, and in the interest, of all mankind. Secondly, the activities of states in exploration and use of outer space shall be carried out in accordance with international law, including the Charter of the UN, in the interest of maintaining international peace and security. Thirdly, outer space and the celestial bodies are not subject to national appropriation by claim of sovereignty, by means of use or occupation or by any other means. Of course, the earth is excluded from the celestial bodies.

It is proposed to advocate that some of the fundamental principles enshrined in the OST already comprise the customary law of outer space and these elite precepts may be acknowledged and propagated as *jus cogens* of space law. The author is conscious that many scholars may not wholly concur with this viewpoint and have reservations about it, yet it is humbly offered for consideration, debate and refinement.

Concept of Customary International Law

Customs are usages of a community or group of persons that are habitually and voluntarily obeyed as a 'conditioned response' under similar circumstances by a large number of the group. Such habitual responses evolve as a group norm, over a period of time, to acquire the force of law to ensure invariable compliance. In other words, "Customary law [is where] established usages…come to be regarded as having an obligatory character…"[6] This is customary law where every member of the group feels impelled to conform and does not want to behave differently or act anti-social lest he be singled out for breach and criticism for being non-conformist even when the statute law is non-existent. Thus, customs are evidence of general practice or consensus accepted as law. Here each member has no 'social contract' or legal obligation to cooperate or abide, yet customary law, though soft law, has a binding force that is unimpeachable and its hold is really strong.[7]

In similar manner, the usages of international relations that become part of consistent state practice and over a period get emulated for adherence by other countries in common habitude of state interaction come to be accepted as customs of international relations. Some jurists have pointed out differences between customs and usages. They insist that the two terms are not synonymous and bear a distinctly different meaning in jurisprudence. To amplify the distinction, customs crystallise "when a clear and continuous habit of doing certain actions has grown up under the aegis of the conviction that these actions are, according to International Law, obligatory or right. On the other hand, international jurists speak of a *usage* when a habit of doing certain actions has grown without there being the conviction that these actions are, according to International Law, obligatory or right."[8]

The point relating to 'conviction' in the obligatory nature of international custom needs amplification. International custom must meet two criteria. First, it must show acceptance of practice, "expressly recognised and accepted," and its consistency of conduct by the states and, secondly, an implicit yet overt belief in its legal validity and compulsive character. This is a psychological element in the establishment of customary law and can be expressed as *opinio juris sive necessitatus*.[9] It indicates a conviction of its legal obligation. However, there is a wealth of state practice that does not usually carry with it a presumption of *opinio juris* yet it could be deemed the settled practice of states.

Further, as these customs get universalised as natural responses under given conditions, these gain the force of normative behaviour with compulsive hold for voluntary adherence. At some stage, the customary rule gets abstracted from the individualised conduct and turns into customary international law that can euphemistically be called "World Discipline for International Relations". The point in time when this process culminates is a matter of fact and not of theory. However, the metamorphosis is, thus, complete and the ultimate in international law is reached.

It follows, therefore, that international law is a dynamic corpus that keeps developing and changing according to variations in relational patterns and mutations in practices. In this process of transformation, new customs supersede older treaties and new treaties may replace older customs. Thus, treaties have over time gradually displaced or codified customary international law like that of the global commons or *jus ad bellum*. Yet vice versa is equally true because treaties are generally deficient in effectiveness for being not binding on non-parties and a majority of these lack universal ratification. Yet, most of the customary international law has sustained with durability.

Customary laws have developed in two ways. First, they "had their origin in the practice of a single [but powerful] state which was able to impose its will until the rule came to be accepted by other states without protest."[10] Many such rules, for example, relate to maritime warfare. The other method relates to "their origin in the voluntary practice of a small group of states, and being found useful and convenient, were gradually accepted by other states until the established practice became a binding rule."[11] Such rules have come into being relating to diplomatic immunity, international commerce and trade relations.

However, for hardening of an abstract rule into a concrete practice and an accepted custom, there are no deductive benchmark parameters in terms of reiterated acts of regular observance or frequency of affirmations of a particular principle to show its general acceptance. The slide from precedent to custom is gradual and subtle, with the distinctive character that the acknowledged practice acquires a manifested recognition of a lawful obligation. It seems germane to say that international law is based on the consent of states, which could be express or tacit. Customary international law carries the *tacit* consent of states that could be implied or expressed in conduct.[12]

It also becomes pertinent to highlight the importance of customary international law by alluding to its honurable reference in the Statute of the International Court of Justice annexed to the Charter of the United Nations. Article 38 in Chapter II of the Statute relating to the competence of the court states *inter alia*, "The Court…shall apply international customs as evidence of a general practice accepted as law." Thus, this Article is generally recognised as a definitive statement of sources of international law. Further, to avoid the possibility of *non liquet*, sub-para (c) has been added after para (b) that explicitly mentions international customs. In a nutshell, international custom is a source of international law with equal importance and equal validity with treaties and pacts *et al.*

OST as Customary International Law

The United Nations was inspired by the great prospects opening up before mankind as a result of man's entry into outer space and recognising the common interest of all mankind in the progress of the exploration and use of outer space for peaceful purposes, and believing that such activities should be carried out for the betterment of mankind and for the benefit of states, irrespective of their degree of economic or scientific development. As a result of this concern about activities relating to outer space, the UN has made tangible efforts to regulate such human activities by establishing a few cardinal principles. This solicitude found expression in the UN General Assembly Resolution 1721 (XVI) of December 20, 1961,[13] that comprised five documents laying down the first set of rules governing outer space. This was followed by Resolution 1802 (XVII) of December 14, 1962, that was adopted unanimously by the states members of the United Nations.

Codified space law came into existence with the Declaration of Legal Principles Governing the Activities of States in the Exploration and Use of Outer Space which was adopted through General Assembly Resolution 1962 (XVIII) of December 13, 1963. This code was reiterated and further elaborated as the Treaty on Principles Governing the Activities of States in the Exploration and Use of Outer Space, including the Moon and other Celestial Bodies.[14] This document continues as a basic statute comprising the fundamental principles of space law and it should be a matter of much gratification that this regime has been adhered to with hardly any intentional aberrations by the space-faring states.

General Assembly Resolutions express the will of the states and were resorted to for simplification of the treaty-making procedure. These are "oral agreements" but are "undoubtedly international agreements subject to the law of treaties." The International Law Commission has also confirmed in its commentary that Oral International Agreements are "a new type of international instrument, which, belonging to the realm of law, may, under concrete circumstances acquire all the characteristics of a binding international instrument."[15] This view has also been reiterated by the Office of Legal Affairs of the UN by pointing out, "In United Nations practice, a 'Declaration' is a formal and solemn instrument, suitable for rare occasions when principles of great and lasting importance are being enunciated..." Therefore, in view of the significance of the declaration, it may be considered to impart, on behalf of the organ adopting it, a strong expectation that members of the international community will abide by it.[16]

Some jurists are of the view that General Assembly (GA) Resolutions are not binding on states in content and law unless the states wish to abide by them. These are also not enforceable because the General Assembly lacks enforcement powers or the right to sanction. Thus, GA Resolutions are platitudinous statements symbolising the sense of the international community on a global issue. But the empirical reality is more potent in impact and these resolutions seem to carry considerable political weight and legal authority. Further, the General Assembly has an option to refer any issue to the Security Council for consideration and, if deemed necessary, to put in place a binding resolution with sanctions.

Moreover, the validity and force of GA Resolutions depend upon their substantive contents, their enduring importance to the world community and the manner in which the resolutions are carried to adoption. It has been the experience that the resolutions relating to governance of space activities and treatment of outer space, including the celestial bodies, have invariably found unanimous support, with hardly any dissenting voices; thus, under the principle of unanimity, the declarations hardly needed any voting procedure. The resolution relating to the Declaration of Legal Principles was adopted by acclamation.[17] This declaration "could not be viewed as a mere recommendation,...it was an international instrument which confirmed and created law on the subject."[18]

The universal appeal attracted by these resolutions and the OST lends them an instant acceptance and the halo of customary law. Another reinforcing point is that this treaty became effective within less than a year. Diplomatic circles are aware of the time consuming processes and dilatory formalities that delay ratifications. The speed of deposit of accessions reflects that there was strong unanimity of opinion on the issue, with hardly any conflicting views to cause delay. As on January 1, 2011, there were 101 parties and 26 signatories to the treaty.[19] This can lead us to a robust belief that the OST has become customary international law.

Concept of *Jus Cogens*

It seems too trite to assert that international law is based on the maxim of *pacta sunt servanda* and has depended on treaties and agreement and the will of the state for compliance or acceptance to operate in conformist behaviour within the comity of state-parties. Starke defines international law "as that body of law which is composed for its greater part of the principles and rules of conduct which states feel themselves bound to observe, and, therefore, do observe in their relations with each other."[20] It is, thus, a code of conduct of equals or the law of nations and would not be deemed as a supra-national law[21]

Over a period of time, through state practice, some of the basic principles get universalised in adherence and crystallise as customary international law which, in turn, requires no specific treaty or ratification from new states to obligate them to accept and honour its existence and participate in its systemic operation. These exist and operate *a priori* for the new states. Thus, law, in a historical perspective "is an expression of customary morality which develops silently and unconsciously from one age to another."[22] The process continues unabated into the future.

With the passage of time, some tenets of the customary international law reach so close to *jus naturale* or Vattel's "natural law of reason" as to be deemed fundamental and universal, thus, being embedded in the human psyche as also reflecting in consistent state practice that these get exalted to the status of *jus cogens*.[23] These are, thus, peremptory norms of general international law that the states are not permitted to denounce or contract out in any pact or treaty. And should it so happen, the concerned provisions can be deemed null and void in concordance with the jurisprudence of the International Court of Justice (ICJ). These are customs that are treated as implied agreements and tend to be binding even without ascension or ratification. "Thus, norms of International Law have the character of *jus dispositivum* or if there exist some norms, having the character of *jus cogens* too from which no derogation is permitted by an agreement *inter partes*."[24] Hence, treaties must not in any manner contradict or conflict with such fundamental principles and precepts.

The principle of *jus cogens* is also enshrined in Article 53 of the Vienna Convention on the Law of Treaties, 1969, and is treated as a peremptory norm of general international law that is accepted and recognised by the international community of the states as a whole. It is a norm from which no compromising detraction is permitted and which can be modified only by a subsequent norm of general international law having the same legal character and normative strength. In other words, *jus cogens* are special principles with the halo of *opinio juris* that prohibit a state from committing internationally wrongful acts. The European Court of Human Rights has also stressed on the International Public Policy aspect of *jus cogens*.

Jus Cogens and State Sovereignty

Another pertinent aspect that deserves consideration is that pedagogues from the modern positivist doctrine of international law assert that the power and competence of states to conclude treaties and pacts is, in principle, unfettered and unlimited. *Jus* is the legitimate creation of the sovereign authority of a state that is not accountable to any superior authority and *jus* cannot be deemed senior to its creator. This view accords with the principle of classical sovereignty expounded by Kant and in the Hobbesian *Leviathan* as well as sovereign equality of states under state practice and rightfully assured under the UN Charter, whereas *jus cogens* as a notion, paradoxically, subordinates the august imperiality of sovereignty. There is, thus, an inherent *contradiction in terminis*.

The traditional doctrine of sovereignty has an inalienable aspect of territoriality over which a sovereign is supreme and wields indisputable authority. This milieu exists on the planet earth and it in no way prevails in outer space and on the celestial bodies which admit of no territorial sovereignty nor national boundaries. If at all, in outer space, the sovereignty of collective humankind holds sway and governance is through the common heritage of mankind. Thus, the dispute with national sovereignty practically vanishes in outer space and on the celestial bodies. Therefore, it would be prudent to shed the obsolete and discardable baggage of the past, and it seems no mission is impossible. Hence, under this changed paradigm, *jus cogens* arise as exalted norms of international law that tend to provide justice to humanity at large. Therefore, as a concept, these command peremptory authority, remain mandatory in operation and admit of no derogation from national sovereignty.[25]

It is also arguable from a practical angle that over the last century, the hardness of sovereignty had been softening with the establishment of the League of Nations after World War I and later with the constitution of the more powerful United Nations. The latter is more or less a supra-national body, with powers to discipline errant sovereign nations and, thus, the basic concept of sovereignty has been badly dented. Contemporary trends indicate that it has been further compromised by the implied powers of the International Court of Justice, establishment of the International Criminal

Court, specialised International Tribunals and other dispute settlement mechanisms, international and inter-governmental organisations as well as non-governmental organisations.[26] These have all demolished the myth of paramountcy of sovereignty and eroded its command. The vision of the phantom of sovereignty is turning hazy and its dominion is waning. And, fortunately, it has happened gradually and voluntarily and not through any coercion by conventions or a revolutionary process. In contrast, the halo of *jus cogens* has been becoming progressively sharper and distinctive.

A dilemma accosts us as to whether *jus cogens* can demote or restrict this inherent authority of the states. Hans Kelsen and Georg Schwazenberger could not bear to denigrade the sanctity of national sovereignty and failed to subscribe to the concept of *jus cogens,* and accorded supremacy to state consent to perpetuate the sway of treaties. However, Hersch Lauterpacht holds a contrary opinion though he reserves uncertainty over the content and scope of *jus cogens*. He asserts that *jus cogens* are necessary for a superior "*Ordre Internationale Publica*" in an unorganised and asymmetrical international society but laments the lack of modalities to identify the genuine from the illusory. He believes that these peremptory norms derive their unique legal authority from two inter-related sources: international morality and general principles of state practice.[27] Thus, this theory of obligations is more concerned with international public order and public policy of international relations. It is wisely said that transcendental morality is tantamount to international public policy.

In fact, sovereignty today is only an emotional hook; its erosion has already started and the myth is bursting. The old belief that sovereignty was an indivisible composite is no longer sacrosanct. It is now considered constitutive, that can be partly surrendered and still be exercised in practice. As a result, *de facto* sovereignty and respect for states remains mostly undisturbed. Legal scholars differ in their ideas, with no settled opinion. For instance, Weil calls *jus cogens* a pathological phenomenon.[28] Despite dissent on the subject, there is a discernible trend emerging over the last three decades to accept *jus cogens* as a matter of necessity in international relations.[29] It remains a moot issue[30] yet deserving of a revisit.

The Soviets, generally, in state practice, disparage customs as a source of international law and impute treaties, signed and acceded to, alone with binding force, while the Soviet Constitution establishes "the principle of observing in good faith the obligations emanating from the generally recognized principles and norms of international law."[31] It would, of course, seem interesting to find Russia (erstwhile USSR), in principle, a strong votary of this concept of *jus cogens*.[32]

Moral Element in *Jus Cogens*

It needs no citation to aver that law is basically amoral and it would be equally useless tautology to prove that *jus cogendum* are not only related to morals but are imbued with them. Thus, the character of both differs which has led a scholar to assert *jus cogens* as juristic illusion.[33] The fallacy here is in empirical observation of state practice and interpretation of almost invariable compliance of *jus cogens*. Of course, there are deviances and aberrations but that proves that a rule of law exists and violations can attract sanctions.

It needs no affirmation that *jus cogens* become acceptable based on pith and substance of the enshrined rule, the spirit of which is deeply rooted in international conscience, thus, deriving their inherent rationale and moral authority. In consequence, *jus cogens* abdicate their dependence on state consent to effectuate compromise on sovereignty. As Vattel puts it, *jus cogens* are a universal natural law of reason based on justice and morality and assuring a juridical order with ethical core values for moral and peaceful coexistence of the members. One can also seek endorsement from Fuller's doctrine of internal morality of law and establishment of detailed criteria—formal and substantive—as desiderata for selection of *jus cogens*.[34]

This takes us to a higher plane of *obligationis erga omnes*, the rules of which are constitutive of the international community and comprise the minimum requirement for durable peace and public order in the global comity. These are essential for the protection of the fundamental interests of international society — that derelictions that disturb the maintenance of peace and security constitute international crimes that can be tried and punished. These are, in other words, *jus cogens*, binding and enforceable rules of mandated normative behaviour and compulsory orderly conduct in international relations. These mostly are *lex lata*.

Another school of thought treats *jus cogens* as *lege ferenda*, implying that *jus cogens* represent an idealist view of law, as it should be, and it develops from *lex lata*, that is, the existing law. The mechanics of development is correct yet it begs a question as to at what stage a rule of law becomes *jus cogens* and what are the qualifying tests of such graduation. Here one can seek good succour from the persuasive criteria evolved by Fuller that categorises selection procedure under formal and substantive sets of desiderata.[35]

The formal criteria comprise, first, that it should be a general and universal principle and not a peculiarised command of limited jurisdiction. Secondly, it should be clear, explicit and unequivocal, with internal consistency. Thirdly, it should be in the public domain and not a secret pact of limited applicability. Fourthly, the rule should have been consistently followed in practice and relatively stable over a period of time. And, lastly, the rule should be feasible in compliance and not demand *ad impossibilia*. Further, it should be with prospective enforcement and not with retrospective effect. These desiderata are more gestural and procedural.

The other set of considerations relates to substantive criteria that concern the structure and content of the rule of law competing for recognition as *jus cogens*. These attributes relate to the quality, content and operability of the principle under evaluation. The criteria are, first, it should be a principle of integrity, fairness and reasonableness, assuring freedom and security. Second, it should be a rule of law prescribing moral equality that is fundamental, mandatory and non-derogable. Third, the state as fiduciary should dutifully act as protector and agent of its nationals and in pursuance of the principle of solicitude, protect the legal interests of the subjects. And, last, it should reflect the dignity of intrinsic value and should be able to fit compatibly within the framework of *jus cogens*.

From the foregoing, it comes out clear that *jus cogens* certainly bear an inherent element of morality and idealist values that can, at times, be controversial in the multi-cultural, multi-more world community. Yet in the metamorphosis of *jus cogens*, their core content is initially derived from *lex lata,* or an antecedent law that was already in force as pact or law. It is its wide appeal and empathetic psyche that pushes it to the realm of *lege ferenda* and, finally, its universal and explicit acceptance in the international community that converts *jus a priori* into exalted *jus cogens*. Nonetheless, the concept of *jus cogens* is laudable and practicable and it is here that the sagacity of humanity should prevail to sift the chaff from the grain.

Fiduciary Aspect of *Jus Cogens*

Of course, many jurists believe that *jus cogens* override other sources of international law, even the Charter of the United Nations. Take, for example, the Fiduciary Theory of *Jus Cogens*[36] where Immanuel Kant discusses the innate right of the child against parents as a familial fiduciary relationship.[37] In analogy, humanity can be placed in parallel as a child in trust, with, say, the United Nations, and *jus cogens* assure its rights. In conformity, international law accepts the superior order of *jus cogens* over the statute law of national sovereigns. In other words, *jus cogens* are the trustees of higher public order to impart natural human justice to mankind.

The fiduciary theory lays great stress on the responsibility and obligation of states. It, thus, has a fiduciary obligation to govern its nationals "in accordance with the principles of integrity, fairness and solicitude as well as to provide equal security under the rule of the law."[38] It embargoes political prohibitions and upholds venerable norms like *jus cogens* that bear endemic peremptory force. "The Fiduciary Theory, thus, offers a principled basis for revitalizing the *jus cogens* concept in International Legal Theory and in the jurisprudence of national and international tribunals."[39]

In consonance, if the state is the fiduciary of the individual subject to its power, then the relationship between the international legal order as a whole and the individual too carries an element of the fiduciary trait, in that the international community may feel enjoined "to act as surrogate guarantor of *jus cogens* if the state flagrantly violates peremptory norms." Thus, the fiduciary principle seeks to vindicate the individual's innate right to be treated as a person with equal dignity under the panoply of fundamental rights guaranteed under the doctrine of universality.

As a corollary, the state as the agent of its nationals, has a solemn duty, on actual or constructive knowledge of the problem or distress, to diligently represent its people and act for their utmost welfare at home, and bodily protection abroad and on their behalf, without specific authorisation. This is a *jus cogen* of international law and the state cannot contract out nor delegate, to, say, military dictators, this essential fiduciary obligation of statehood (like policing for law and order) and its allied duty to ensure rule of law and guarantee of equal security to all subjects. *Jus cogens*, thus, impose a specific duty on the state to safeguard the dignity of its nationals as a multifaceted and overarching obligation on the state.

Nevertheless, the state is a powerful entity with wide discretion and ample immunity while the individual as the beneficiary is in peculiarly vulnerable predicament where his dignity and welfare can be fouled and compromised by the state-trustee with gross impunity. Therefore, the hallmark of the fiduciary duty of the trustee is unflinching loyalty towards the ward under trust, consistent with the principles of equality, non-discrimination and self-determination, with due freedom and dignity. *Jus cogens* exactly perform this function to guarantee such established rights in the international comity. Here *jus* is perceived in its surrogate function in the legal dynamics of the state-subject fiduciary relationship.

Jus Cogens of Space Law

An effort has been made to cull out some basic principles of space law that bear universal appeal and carry unanimous acceptance among the comity of nations as peremptory norms to command invariable adherence. These venerable precepts can be extolled as *Jus Cogens* of Space Law. It is, however, conceded that this list is neither complete nor comprehensive and is open to debate for refining the concept and honing its nuances. It is, thus, changeable and perfectible. In order to correctly grasp the import and comprehend the niceties of this principle, it is proposed to illustrate a few provisions of the space law, emphasising the precepts that are based on the solid foundations of law, treaty and state practice.

Outer Space as Province of Mankind

The Outer Space Treaty enshrines a laudable principle that outer space and the celestial bodies "shall be the province of mankind,"[40] implying that the entire universe belongs to humanity as a whole and is, thus, *res communis*. This notion is a variant of the famous concept of Common Heritage of Mankind first incorporated in the Law of the Sea and later adopted into the Legal Regime of the Global Commons.

In general connotation, this conceptual phrase has two parts: first, the word 'province' emphasises historical distinctness and different characteristics of the expanse of outer space and discrete territorial traits of the celestial bodies that comprise the 'space system;' a new frontier that is strategically and politically divergent from the planet earth. In other words, it is province of the empire of the universe minus the planet earth. The second word 'mankind' implies humanity. But both terms are generic and vague in what they encompass, their future composition and for variations of their comprehension. Further, these have no legal entity or legal capacity as subjects of international law. In the treaty, the basics of the concept of 'province of mankind' do not relate to legality but are stated as a precept that activities in outer space "shall be carried out for the benefit and in the interest of all countries…" It is, thus, a primary principle of space jurisprudence.

As a corollary to the above, follows a provision in the OST stating, "Outer space, including the Moon and other celestial bodies, is not subject to national appropriation by claim of sovereignty, by means of use or occupation, or by any other means."[41] The treaty Article prohibits all states of the international comity from asserting national ownership or any proprietary rights in outer space or on any celestial body for any reason or by any means. This mandates that outer space and the celestial bodies are to be regarded as *res nullius* or better still, *res communis* or *res publica*.

Outer space "shall be the province of all mankind"[42] shifts the emphasis from the traditional postulate of national sovereignty to international cooperation, with community rights for the common good, thus, highlighting the underlying principle that there are areas where the common interests of mankind must be served and given primacy. This clause concedes the possibility of a conflict of ideology or clash of national interests in space operations but dispels "any such spectre to seek a common vision of their future relations in a newly accessible environment."[43] This principle strengthens the sense of the international community with *de facto* respect to other countries to create common interest and encourage collective security for the sake of mankind.

The Moon Treaty also, in Article 4, directs, "Exploration and use of the Moon shall be the province of all mankind and shall be carried out for the benefit and in the interest of all countries… to promote higher standards of living…"[44] It further adds, "States-Parties shall be guided by the principle of cooperation and mutual assistance in all their activities concerning exploration and use of the Moon."[45] It, thus, brings in sharp relief the concept of community ownership of outer space as a 'province of mankind.' This is in contrast to the rules of territoriality under international law on the planet earth. But no dissent has been vocalised to dispute this norm of space law. Thus, the unanimity displayed in its acceptance elevates it to the podium of *jus cogens*.

Freedom of Access to Outer Space
The Outer Space Treaty offers unique freedom to all states, irrespective of their technological advancement, for activities in outer space and on the celestial bodies that shall be without restrictive frontiers and without national boundaries. The freedom is universal and futuristic, without let or hindrance, and not related to existing threshold of capabilities for such activities. The treaty emphatically states, "Outer space, including the Moon and other celestial bodies, shall be free for exploration and use by all States without discrimination of any kind, on a basis of equality and in accordance with international law, and there shall be free access to all areas of celestial bodies."[46] Humankind has never enjoyed such freedom and the principle enshrined is, indeed, laudable.

However, this freedom is not absolute and carries with it reasonable restrictions and corresponding duties. First, "…the States shall be guided by the principle of cooperation and mutual assistance and shall conduct all their activities in outer space, including the Moon and other celestial bodies with due regard to the corresponding interests of all other States-Parties to the Treaty."[47] This provision sounds basic and a true *jus naturale*.

The next restraint binds that states "shall carry on activities in the exploration and use of outer space, including the Moon and other celestial bodies, in accordance with international law, including the Charter of the United Nations, in the interest of maintaining international peace and security and promoting international cooperation and understanding."[48] This clause has an inherent flaw because international law acknowledges sovereignty and annexation of territory by different methods and means on the planet earth. This delinquency has been rectified by an explicit and express assertion that, "Outer space, including the Moon and other celestial bodies, is not subject to national appropriation by claim of sovereignty, by means of use or occupation, or by any other means."[49] Humankind has traditionally espoused sovereignty and has lived with state boundaries; therefore, the principle of community ownership of outer space and celestial bodies without frontiers appears naive and novel. Nevertheless, it is a great departure from the age-old mindset of dividing territory and constitutes a quantum leap into a new and higher level of legal regimen that has been accepted universally. It, thus, certainly qualifies as a *jus cogen*.

Another reasonable restriction relates to the international responsibility of the states towards activities in outer space and on the celestial bodies carried on by governmental agencies or its nationals or by non-governmental entities. The states shall assure "...that national activities are carried out in conformity with the provisions set forth in the present Treaty."[50] A corollary to this attaches international liability to the state for any damage caused by its activities to "another State-Party to the Treaty or its natural and juridical persons by such [launched] objects or its component parts on the Earth, in air space or in outer space, including the Moon and other celestial bodies."[51] Compensation for damage caused is an established legal precept under torts, contracts and civil laws and, hence, vindicates freedom of outer space as a *jus cogen*.

There is assurance of an additional "freedom of scientific investigation in outer space, including the Moon and other celestial bodies, and States shall facilitate and encourage international cooperation in such investigation."[52]

This facilitation is also circumscribed by correlative duties and obligations. The states shall undertake experimentation and "...pursue studies of outer space, including the Moon and other celestial bodies, and conduct exploration of them so as to avoid harmful contamination and also adverse changes in the environment of the Earth resulting from the introduction of the extra-terrestrial matter and, where necessary, shall adopt appropriate measures for this purpose."[53]

In case, a state "...has reason to believe that an activity or experiment planned by it or its nationals in outer space, including the Moon and other celestial bodies, would cause potentially harmful interference with the activities of other States-Parties in the peaceful exploration and use of outer space, including the Moon and other celestial bodies, it shall undertake appropriate international consultations before proceeding with any such activity or experiment."[54] Also, if a state has reason to believe that an activity or experiment planned by another state would cause similar potential harmful interference or adverse effect, it "...may request consultation concerning the activity or experiment."[55] The freedom is mutual and reciprocal, so is the correlative obligation. Of course, the mandate is, indeed, fundamental to be considered a *jus cogen*.

State Responsibility to Humanity

That states shall bear international responsibility to state-parties as also responsibility to humanity at large, for national activities and those of its non-governmental entities and individuals, is another mandate of space law. The principle of state responsibility is a classical doctrine of international law and has been adhered to for centuries: and new connotations of state responsibility have gradually evolved with progressive times and the changing international milieu.[56] State responsibility is a correlative of international obligation and this concept has now been embedded in space law with a 'customary' legal mandate and a higher normative value. Nevertheless, its compliance is equally obligatory and attendant with sanctions.

This doctrine is expected "to serve as a specific instrument of legal regulation in international relations and stimulate the functioning of international law."[57] In general terms, state responsibility refers to the legal consequences for action of its nationals, including executive organs of governments and natural persons, as subject to international law, that follow upon violation or a *delictum* or an act of commission or omission relating to any international legal obligation. It may be added for clarity that the state responsibility extends to harmful consequences of even legitimate activities by its nationals. Thus, any such failure or detrimental effect, in turn, sets up legal liability *qua* aggrieved nationals of another state subject to the basic rule that all domestic options of protection and remedies must first be exhausted.

One is impelled to allude to another relevant aspect of state responsibility that can be sublimated to *erga omnes* and this rule has since been recognised in customary international law that *pari passu* becomes applicable to contemporary space law. The legal force of this particular obligation that is owed by the states to the international community as a whole has been identified and obliquely highlighted by the International Court of Justice in the Barcelona traction case[58] among others. As a result, this humanitarian duty of the state towards humanity at large has been accepted universally and has got deep-rooted in state practice. In fact, state responsibility *erga omnes* has been elevated to the status of a *jus cogen* of space law and violations of *erga omnes* obligations and peremptory norms may be punishable by any state under the universality principle.[59] This *jus cogen* can also draw substantive support from the Vienna Convention.[60]

In order to correctly grasp the relevance and to comprehend the nuances of this principle, it is proposed to illustrate from the space law emphasising this precept. The OST ordains that "States-Parties to the Treaty shall bear international responsibility for national activities in outer space, including the Moon and other celestial bodies, whether such activities are carried on by governmental agencies or by non-governmental entities, and for assuring that national activities are carried out in conformity with the provisions set forth in the present Treaty."[61]

It is also pertinent to make a mention of the responsibility, that states "...shall conduct all their activities in outer space, including the Moon and other celestial bodies, with due regard to the corresponding interests of all other States-Parties to the Treaty." Further, they shall "... pursue studies of outer space...and conduct exploration of them so as to avoid their harmful contamination and also adverse changes in the environment of the Earth resulting from the introduction of extra-terrestrial matter and, where necessary, shall adopt appropriate measures for this purpose."[62] In negative terms, it is the states' responsibility not to despoil or pollute the medium of outer space and the environment of the celestial bodies, and its adverse fallout on the earth.

The treaty attributes responsibility on the states for acts or omissions for activities in outer space under their control and jurisdiction and their effect and impact on the earth. The attribution of responsibility leads to liability as *vinculum juris* for an act or omission which violates a right or fails in its obligation established by the rule of law and, in consequence, causes an injury. The *injuria* has to be ultimately monetarily compensated per *restitutio in integrum*. The liability in such cases is absolute and indefensible. The sagacity of this provision is unimpeachable and this principle deserves to be elevated as a *jus cogen* of space law.

International Cooperation as Cardinal Principle

The OST is replete with references to cooperation between states and the relevant provisions are meaningful and binding. To begin with, the Preamble to the treaty optimistically exhorts states-parties, "...desiring to contribute to broad international cooperation in the scientific as well as the legal aspects of the exploration and use of outer space for peaceful purposes." It further believes "...that such cooperation will contribute to the development of mutual understanding and to the strengthening of friendly relations between States and peoples."[63] The hope is sincere and the intention permeates all through the treaty.

Article I of the treaty mentions, while referring to the freedom of scientific investigation in outer space and the celestial bodies that "...States shall facilitate and encourage international cooperation in such investigation." Again Article III, while permitting space activities urges for "...maintaining international peace and security and promoting international cooperation and understanding." Further, Article X endorses international cooperation to afford "...an opportunity to observe the flight of space objects launched by ... States." Extending the principle, Article XI commands to "promote international cooperation in the peaceful exploration and use of outer space...including the Moon and other celestial bodies,...to inform...of the nature, conduct, locations and results of such activities."[64]

Further, astronauts have been exalted in status and accorded the privilege of being the envoys of mankind in outer space with the intent to involve all states into this notional concept and elicit their unflinching cooperation for supporting space activities.[65] Similarly, the very basis of the Agreement on the Rescue of Astronauts, the Return of Astronauts and the Return of Objects Launched into Outer Space, 1968, is essentially for cooperation and as stated in the Preamble it has been "prompted by sentiments of humanity." The tenor of the emphasis is clear and explicit.

The Moon Treaty[66] too in its very Preamble, is, "determined to promote on the basis of equality, the further development of cooperation among States in the exploration and use of the Moon and other celestial bodies." Again in Article 2, it maintains, "All activities on the Moon...shall be carried out...in the interest of maintaining international peace and security and promoting international cooperation and mutual understanding and with due regard to the corresponding interests of all other States-Parties." The tone, it will be noticed, is conciliatory and cooperative.

Article 4 directs, "Exploration and use of the Moon shall be the province of all mankind and shall be carried out for the benefit and in the interest of all countries... to promote higher standards of living..."[67] It further adds, "States-Parties shall be guided by the principle of cooperation and mutual assistance in all their activities concerning exploration and use of the Moon. International cooperation in pursuance to this Agreement should be as wide as possible."[68] The stress on the importance of international cooperation among the comity of nations, and in particular space-faring countries, is unmistakable

Such and similar expressions of advisory nature or as pseudo-obligations are copiously found in other instruments like the Agreement on the Rescue of Astronauts[69] and Convention on Registration of Objects Launched into Outer Space, 1975. However, the corpus of space law, besides treaties and agreements, comprises several guidelines and in-principle documents and codes of conduct regulating space activities and governing its uses. These are contained in resolutions adopted by the UN General Assembly and are based on an appeal to maintain public order in the respective field of activity and urge for voluntary cooperation to best satisfy individual needs and simultaneously reap optimum benefits for mankind.

A significant example could be the principles for regulating activities in the field of international direct television broadcasting by satellites.[70] This area involves allotment of slots for geo-synchronous satellites and sharing of telecommunication spectrum and radio frequency allotments which "should be carried out in a manner compatible with the sovereign rights of States, including the principle of non-intervention as well as the [with] the right of everyone…"[71] Further, paragraphs B, H and I of the document also encourage the same principle. In fact, paragraph D is fully devoted to international cooperation and urges "[s]pecial consideration…to the needs of the developing countries… for the purpose of accelerating their national development."[72] An altruistic attitude is clearly recommended.

Other sibling document enunciating international cooperation is the Principles Relating to Remote Sensing of the Earth from Space[73] which covers a very sensitive issue so very delicately poised in the corpus of space law. This aspect comes in sharp relief in the paragraphs stating Principles V, VIII and XIII. For illustration, Principle V states that "remote sensing activities shall promote international cooperation…on equitable and mutually acceptable terms."[74] Another pertinent example is that of the UN Guidelines for Mitigation of Space Debris, 2007, that lament the quantum of existing accumulation of detritus, particularly in the lower reaches of outer space and the almost exponential addition to the same every decade. In remediation thereof, it specifically solicits cooperation among states and coordination of technological resources for reduction in redundant satellites, scavenging of space litter, avoidance of incidental debris and conscious effort to shun advertent creation of junk in outer space.[75] Cooperation will ensure that outer space operations are safe, viable and sustainable till into the distant future.

The Declaration on International Cooperation[76] adopted by the UN General Assembly is solely devoted to promoting and fostering the cult of international cooperation among states. The declaration "recognizes the growing significance of international cooperation among States and between States and international organisations..." for "efficient collaboration ...for the mutual benefit and in the interest of all parties...taking into particular account the needs of developing countries." The sincere concern of the UN for cooperation and amity among nations for their own betterment and welfare as well as for conflict avoidance and dispute resolution in outer space activities becomes amply evident and should be taken as an obligatory plea rather than a mere exhortative or impassioned appeal.

And lastly, the European Union (EU) Code of Conduct[77] encourages cooperation, consultation and mutual assistance to "seek solutions based on an equitable balance of interests." This accepts and imparts a new understanding of sovereignty that is at variance from the one imbedded in traditional international law. The EU has urged nations worldwide to join and adhere to this code to make its acceptance almost universal. The US is also preparing a similar Code of Conduct for Outer Space.

One can also solicit support from the views of Goedhuis that in meeting the varied challenges of the space age, man has been able to cavort and combine the forces of the social complex which provide a realisation of greater world interdependence because of the limitations of technology and the ultra-hazardous domain of outer space and, thus, necessitate cooperation. Inclusivity of all states, technically capable or still struggling, is inalienable and integral to the order of outer space. Thus, an important feature of space law reflects the gradually transforming structure and reveals a process to detoxify international relations of the phantom of sovereignty and highlights recognition of the compulsion of international cooperation in the field of outer space.[78]

It is, thus, held by many scholars that space law contains stronger cooperative duties and obligations to collaboration than general international law. Rudiger Wolfrum has particularly stressed that this principle marks a significant break and a conceptual transition away from the traditional international law of coexistence to a new law of cooperation.[79] Similarly, Rudolf Dolzer holds that the structure of space law is based on active cooperation and mutual assistance complemented by specialised duties towards activities in, and relating to, outer space. This reflects the concept of obligation of assistance with voluntary spontaneity and in the form of reciprocity.[80]

From the foregoing analysis, it gets amply substantiated that international cooperation is the cornerstone of space law for the betterment of mankind and improvement in the quality of life on the earth. International cooperation is not a peripheral issue but a strong strand that runs through the regime of space law and becomes the common denominator of treaties and agreements, principles and guidelines. Therefore, this concept has been fully imbibed and internalised and has, thus, become integral to its functioning and the normative behaviour of all states. Thus, this principle undoubtedly assumes the status of a *jus cogen*.

An Appraisal

Space law is still young and growing but has matured faster in its sagacity and acceptance by the comity of nations. Most of this regime comprises soft law in terms of principles and guidelines but the other part comprises treaties and agreements. The legal regime of outer space, though novel in several respects, has gained wide acceptance rather fast. The Outer Space Treaty is unique in its doctrinal contents and socialistic approach. Despite drastic changes, the pervasive adherence achieved so far certainly appears beyond coincidence and is advertant and volitional by the states. It was initiated as a UN General Assembly Resolution and was adopted by acclamation. The unanimity on the resolution and the speed of ratifications indicate its spontaneous acceptance and add to its repute and effectiveness. The binding and obligatory nature of the mandate is indisputable and seems to have acquired full and total customary force. As a result, the OST has come to be euphemistically called the *Grundnorm* of Outer Space and generally hailed as Customary Space Law.

Some of the provisions in the framework of the Outer Space Treaty are so basic and fundamental that these represent cherished ideals which come close to *jus naturale*. These are, therefore, deemed implied agreements that are obligatory in nature and require no specific ratification.[81] A tentative selection of *jus cogens*, here, covers only four mandates. The first lofty ideal treats outer space as a province of mankind which is neither open to appropriation by sovereignty nor divisible by borders. It is *res* of mankind for use by mankind and for the ultimate welfare and benefit of mankind. The wisdom of this postulate has empirically proven itself in other domains and it deserves the status of a *jus cogen* of space law.

The second fundamental relates to freedom of access to any and every part of outer space, without discrimination, let or hindrance, for exploration and use of outer space and the celestial bodies. It is, of course, axiomatic that every right has a correlative duty to reciprocally assure the same freedom to others in equal measure. Therefore, the instant freedom is subject to the reasonable restriction that states shall conduct all their activities in outer space, including the moon and other celestial bodies, with due regard to the corresponding interests of all other states. Freedom is a laudable concept of eternal relevance and fit to ascend as a *jus cogen* of space law.

The third universal doctrine of space law relates to the international responsibility of states for the consequences of their activities, whether by governmental agencies or non-governmental entities or juridical nationals, and liability for any damage caused as a result of the conduct of such activities. The liability is absolute and indefensible. This, therefore, follows the established principles of tort, contract and civil law that are time-honoured precepts and deserve the status of *jus cogens* of space law.

The last but not the least in importance is the principle of international cooperation that forms the cornerstone of space law for the betterment of mankind and improvement in the quality of life on the earth. It is certainly not a peripheral issue but a strong strand that runs through the paradigm and becomes the common denominator of treaties and agreements, principles and guidelines. Therefore, this concept has been fully imbibed and internalised by this nascent branch of international law and space jurisprudence that has, thus, become integral to its functioning as the normative behaviour of all states.

Jus cogens of space law act as peremptory norms with supervisory status to regulate inter-state relations and other conduct with international impact in outer space so as to command universal obedience and strict compliance, where violations will attract collective censure and sanctions.

No wonder, aberrations have been rather few and minor in nature whereas states have demonstrated judicious restraint to eschew escalations of conflict even in events and occurrences with such a potential. One can, therefore, optimistically and confidently accept that *jus cogens* satisfy the need for a predictable, transparent, flexible and futuristic legal framework for outer space and the celestial bodies.

It, therefore, appears that the above selected four precepts have transformed into universal and fundamental principles of customary space law to get elevated and hallowed as *jus cogens* of space law. These usher in public order of natural law and effectuate a wider normative agenda for international law with rules integral to inter-state relations. This concept can in tandem draw substantive support from the Vienna Convention[82] so as to be able to dispense true justice *ex aequo et bono* to humanity at large. This is a sure harbinger of a healthy and progressive trend that solicits wide support from like-minded legal scholars.

Notes

1. Anthony D'Amato,. "International Law as an Autopoietic System", in Rudiger Wolfrum and Volker Robens, eds., *Developments of International Law in Treaty Making* (Berlin, 2005), pp. 335-399.
2. For example, analogous regimes of Antarctica and the high seas and aspects of international law like state responsibility, state liability *et al.*
3. Details of these instruments have been cited whenever relevant.
4. Treaty on Principles Governing the Activities of States in the Exploration and Use of Outer Space, including the Moon and Other Celestial Bodies, 1967 (in short, OST).
5. Detlev Wolter, *Common Security in Outer Space and International Law*, Geneva, (UN Institute of Disarmament Research, 2005), p.111. *UNIDIR/2005/29.*
6. Charles G. Fenwick, *International Law* (Bombay: Vakils, Feffer and Simons Private Ltd, 1965), p. 88. Words in parenthesis added for clarity.
7. For a detailed and erudite discussion on the subject, refer Eric A. Posner, *Law and Social Norms* (Delhi: Universal Publishing Company, 2009), First Indian Reprint, pp. 4 ff.
8. *Oppenheim's International Law*, Vol. I, Peace, Seventh Edition by H. Lauterpacht (Orient Longmans Ltd, 1952). p. 25.
9. In short, *opinio juris*.

10. Fenwick, n. 6, p. 88. Words in parenthesis added.
11. Ibid.
12. n. 8, p. 24.
13. This resolution was passed without a vote in the General Assembly.
14. This treaty, referred to as the OST in short, was adopted through General Assembly Resolution 2222 (XXI) on December 19, 1966. The treaty was opened for signatures on January 27, 1967, and entered into force on October 10, 1967.
15. Manfred Lachs, "The Law-Making Process for Outer Space", in Edward McWhinney and Martin Bradley, *New Frontiers in Space Law* (NY: AW Sijthoff, 1969),pp. 18-19. Very few declarations like the Universal Declaration of Human Rights or International Covenant on Civil and Political Rights *et al* have been adopted in this manner.
16. Ibid.
17. Adopted on December 13, 1963. Ogunsola O. Ogunbanwo, *International Law and Outer Space Activities* (The Hague: Martinus Nijhoff, 1975), p. 14.
18. Lachs, n. 15, p. 22.
19. UN Website accessed on June 6, 2012.
20. *Starke's International Law*, Eleventh Edition by I. A. Shearer (London, 1994), p.3.
21. Charles G. Fenwick, *International Law*, Second Indian Reprint (Bombay, 1967), pp. 56-57.
22. Benjamin N. Cardozo, *The Nature of Judicial Process* (New Havens, 1921), pp. 104-105.
23. A Latin maxim meaning "Compelling or Strong Law."
24. Alfred Verdross, "*Jus Dispositivum* and *Jus Cogens* in International Law," 60 *Am. J. Int'l L.* (1966), p. 55.
25. The concept of *jus cogens* includes crimes such as slavery, torture, racial discrimination, murder and the like.
26. Rudiger Wolfrum in Introduction to Rudiger Wolfrum and Volker Roben, eds. *Development of International Law in Treaty Making* (Springer, 2005), pp. 11-12.
27. Dinah Shelton, "Normative Hierarchy in International Law," 100 *Am J Int'l L* (2006), p. 336.
28. Prosper Weil, "Towards Relative Normativity in International Law," 77 *AM j. Int'l L*, (1983), p. 416.
29. Shelton, n. 27, pp. 292, 323.
30. Georg Schwarzenberger, "International *Jus Cogens*?" 43 *Texas Law Review* (1965).
31. *The Constitution of the USSR, 1977*, Article 29.
32. Michael Akehurst, *A Modern Introduction to International Law*, Third Edition, (London, Fourth Impression 1980), p. 46.
33. M. Koskennieme, *The Politics of International Law* (1990).

34. Lon L. Fuller, *The Morality of Law* (1969), p. 33.
35. Ibid.
36. Evans J. Criddle and Evan Fox-Decent, "A Fiduciary Theory of *Jus Cogens*," *Yale Journal of International Law,* Vol. 34, pp. 331-387.
37. Immanuel Kant, *The Doctrine of Right* (Cambridge University Press, 1991).
38. Shelton, n.27, p. 333.
39. Ibid.
40. Treaty on Principles Governing the Activities of States in the Exploration and Use of Outer Space, including the Moon and Other Celestial Bodies, 1967 (in short OST), Article I.
41. Treaty on Principles Governing the Activities of States in the Exploration and Use of Outer Space, including the Moon and Other Celestial Bodies, 1967, Article II.
42. Ibid., Article I, para 1.
43. Wolter, n. 5, p.85. *29.*
44. Ibid., Article 4 (1).
45. Ibid., Article 4 (2).
46. n. 41.
47. Ibid., Article IX.
48. Ibid., Article III.
49. Ibid., Article II.
50. Ibid., Article VI.
51. Ibid., Article VII.
52. Ibid., Article I.
53. Ibid., Article IX.
54. Ibid., Article IX.
55. Ibid.
56. The new concepts of state responsibility relate to, for example, war and aggression, coercion of minorities, denial of freedom by colonial powers and are now extended to international and inter-governmental organisations.
57. G. I. Tunkin, *International Law* (Moscow: Progress Publishers English translation, 1986), p. 223.
58. Barcelona Traction, Light and Power Co. *(Belgium v. Spain)* 1970 ICJ 3, 32 (February 5, 1970).
59. Oscar Schachter, *International Law in Theory and Practice* (1985), p.264.
60. *The Vienna Convention of the Law of Treaties*, Article 53.
61. n. 41.
62. Ibid.
63. Ibid., the preambular paragraphs.
64. Ibid.

65. Ibid., Article V. For a detailed analysis, refer G. S. Sachdeva, "Astronauts as Envoys of Mankind in Outer Space: Resolution of a Dilemma", *Asian Journal of Air and Space Law*, 1 AJASL (2011) pp. 1-22.
66. Agreement Governing the Activities of States on the Moon and other Celestial Bodies, 1979.
67. Ibid., Article 4 (1).
68. Ibid., Article 4 (2).
69. Agreement on the Rescue of Astronauts, the Return of Astronauts and the Return of Objects Launched into the Outer Space, 1968.
70. Principles Governing the Use by States of Artificial Satellites for International Direct Television Broadcasting, 1982. UN General Assembly Resolution 37/92 (annex), adopted on December 10, 1982.
71. Ibid., Paragraph A--Purposes and Objectives of the Document. Parenthesis added.
72. Ibid., Paragraph D—International Cooperation.
73. UN General Assembly Resolution 41/65 adopted on December 3,1986.
74. Ibid.
75. Like testing of Anti-Satellite (ASAT) missiles in outer space by China and the US in 2007 and 2008 respectively.
76. Declaration on International Cooperation in the Exploration and Use of Outer Space for the Benefit and in the Interest of all States Taking into Particular Account the Needs of Developing Countries, 1996. UN Document Supplement No. 20 (A/51/20).
77. The European Union Code of Conduct for Space Activity, October 2010. Refer Michael Listner, "An Update on the Proposed European Code of Conduct", *The Space Review*, August 08, 2011.
78. D. Goedhuis, "An Evaluation of the Leading Principles of the Outer Space Treaty of January 27, 1967," *NTIR*, Vol 15, 1968, p. 40. Also refer Fenwick, n. 6, p. 110.
79. Ibid., p. 25, Rudiger Wolfrum, "The Problems of Limitation and Prohibition of Military Uses of Outer Space," *ZaoRV*, Vol 44, 1984. Also refer R. Wolfrum, "Common Heritage of Mankind," *Encyclopaedia of Public International Law*, Vol 11, 1989, p.67.
80. R. Dolzer, "International Cooperation in Outer Space", *ZaöRV*, Vol. 45, 1985, p. 527. Also refer n. 22, p. 92.
81. Benjamin N. Cardozo, *The Nature of Judicial Process* (New Haven, 1921), pp. 104-105.
82. The Vienna Convention of the Law of Treaties, Article 53.

2

Astronauts as Envoys of Mankind in Outer Space: Resolution of a Dilemma*

The Outer Space Treaty accords astronauts the status of envoys of mankind in outer space, should they come in contact with sentient life in the cosmos. This is a coveted diplomatic privilege that exalts the astronaut above native nationality. The idea created a sense of obligation for all nations in the rescue and protection of crew in distress. It continues to be relevant and meaningful, and deserves to be espoused.

However, from the very beginning, its legality was under a shadow for various reasons like procedural formalities of the issues of diplomatic warrant and accreditation, the nebulous identity of the sovereign authority it sought to represent and the legal status of the collective of mankind as a juridical entity, *et al*. These still remain moot points though relegated to blissful oblivion. International instruments relating to outer space formulated subsequent to the OST have cleverly circumvented this controversial issue.

With advanced space technology facilitating bigger space vehicles and space stations, the concept has been beset with practical difficulties. First, the definition of astronaut is at variance in different space-faring countries. Secondly, with multiple crews on board and each one qualified as an astronaut and, thus, eligible to be an envoy of mankind, a serious problem arises because the law is silent on the criteria for the selection of one. Thirdly, there arises the problem of multinationality crews, all at the same time, and the same place. Considering seniority as the principal criterion, their *inter se* seniority poses ticklish issues endemic of diplomatic discord. And, lastly, with no prescribed protocol or procedure for the designation

---

* This is a revised and enlarged version of an article with the same title from the *Asian Journal of Air and Space Law*, Vol. I, No. 1, January-June 2011, pp. 1-22. Permission of the publishers obtained.

of one particular astronaut as envoy, any one of them could, in such a contingency, react out of turn, under a mistaken belief or a bloated ego or wanton mischief. This could really embarrass humankind.

The dilemma is obvious and pertinent, and seeks an urgent solution. It has been proposed here that the commander of the space vehicle or space station or the space habitat should be so designated to act as the envoy of mankind in outer space, should such a contingency be encountered. For this purpose, the United Nations should take on the evolving responsibility and formulate appropriate rules of conduct.

Introduction

Folk customs worldwide impel societies to honour its heroes for their victories or valour. This tradition has been followed from prehistoric times by almost all civilisations and is still being preserved and perpetuated.. Nations have tended to exalt these "heroes" above the ordinary citizens and accorded them a haloed personality. In this context, the "astronauts" are esteemed heroes of the modern times and tend to carry an aura of significant achievement. Even as per the Outer Space Treaty,[1] they are regarded as "envoys of mankind in outer space,"[2] which implies a status deemed well above the mere nationality of their parent country or that accorded as per diplomatic protocols of international relations.

The euphoria associated with the orbital around the earth by Cosmonaut Yuri Gagarin or the landing on the moon by Astronauts Neil Armstrong and John Glenn has gradually worn off and their halo dimmed by succeeding adventures in outer space like space walks and subsequent deep probe missions that have surpassed these in heroics and hazards. Therefore, it seems just the right time to climb down to reality and assess the current scenario in terms of the number of astronauts in space at any point in time and their *per se* seniority vis-a-vis different nationalities. Besides, we need to factor in the developments of technology that make space transportation of many tourists in a single space vehicle a viable possibility of the near future. It also needs no clairvoyance to visualise smart robonauts[3] working alongside other crew members, with hardly much to distinguish or discriminate them in the eyes of aliens.

State-of-the-art space vehicles like shuttles and commercial space ships can accommodate multiple crews and multiple tourists respectively. Further, with growing space cooperation and reciprocity in activities among states, the crews are increasingly becoming multinational. Thus, the circumstantial presence of multiple crews from different nationalities, all of who are by definition astronauts, creates a multitude of envoys. It, thus, makes for an embarrassing situation for any one member to act as the envoy of mankind, out of turn, on his own and without being so designated by any appropriate authority. The simulated problem is pregnant with one-upmanship, patent dispute and outright embarrassment. Hence, there is an urgent necessity to establish a protocol for the accreditation of one astronaut as the rightful and representative envoy of mankind under diverse situations. This is a grey area in the OST and stares starkly for immediate resolution.

In this chapter, an attempt has been made to probe into the various sources of law — international conventions, multilateral treaties, pacts and agreements, national statutes as also customary practices and diplomatic protocols of civilised nations to cull out the "status of astronauts" that seems legally tenable, universally acceptable and can be honourably accorded. Broadly speaking, all those who have travelled into outer space are astronauts and contemporary reality reveals that the crews of space ships are plural in numbers and of multinational origin, with responsibilities varying in space disciplines and roles tapering in importance while on-board for a mission. To this aggregate of the dilemma can be added the genre of fare-paying passengers and the class of robonauts, also deemed as astronauts. In consequence, the apparent multiplicity of issues gets compounded and the problem comes in clear relief.

Therefore, the question arises whether all of them, severally or collectively, and at the same time, are envoys of mankind in outer space or only one of them should rightly represent mankind as its accredited ambassador. The basic problem is their horizontal cluster, and vertical selection is difficult considering the implications of nationality or *inter se* seniority of the crew members or may be the power of the pelf of the space tourists. Any choice would be open to criticism. Thus, it may tantamount to stating the obvious that the matter seems controversial and would involve subjective opinion and judgmental interpretation. In all humility, a sincere effort has been made to keep the focus as objective as possible.

Definition of Astronaut

The Necessity

Astronaut is an oft-used word in space law and literature but nowhere have the attributes of an astronaut been described for proper characterisation and identification of such a *persona*. Surprisingly, the common notion, though ambiguous and varied, is restrictive in scope and inclusivity. To illustrate on aviation analogy, one wonders whether only the pilot or navigator or the commander of the space ship is the astronaut and, thus, eligible to be an envoy or whether his engineering crew and other mates involved in control, in-flight maintenance, navigation, communication and other allied operations also qualify "to be called astronauts."

One can extend the penumbra of doubt to consider if the accompanying scientists, experimenters and observers too can be brought within the ambit of the generic term of astronauts or be termed as scientists or experts. Liberal definitions yield no benefit either to the defined term or the law. It appears pertinent to allude here to the Moon Treaty that requires that state-parties "shall regard any person on the Moon [and on celestial bodies] as an astronaut"[4] within the meaning of Article V of the Outer Space Treaty "...and as part of the personnel of a space craft"[5] within the meaning of the Agreement on the Rescue of Astronauts, etc. Presumably, such connotation has been accorded for ensuring personal safety and for help in distress.

And, lastly, what about the fare-paying space visitors, space tourists and space fliers who may as well qualify, technically on definition, to be astronauts but happen to travel in the space vehicle for pleasure alone or may be on a personal project but with no officially assigned duty while on board. And, lastly, whether the newly introduced robonauts, *in strictu sensu*, are astronauts as commonly understood and appreciated because they have been touted to be crew members.[6] Loose euphemisms do not qualify for legal vocabulary.

The doubts are valid because space law is a nascent branch of international law and has not yet crystallised fully. Therefore, scientific and juristic elaboration of its glossary has not been undertaken with the zeal and scholarship it merits. Needless to say that lack of, or ambiguities in, definitions have, in state practice and relational experience, caused much avoidable controversy and infructuous debate. Hence, there is an urgent need to benchmark the characteristics and define the traits of an astronaut as a template in the legal lexicon.

'Astronaut' is an actively used term in space law and its description must be crystal clear so that we all understand this *ad idem*. To venture into this is a daunting task as this term has been used variably in international legal treaties and national space legislations. Further, it would be simplistic to presume that equivalent terms, used by different space roving countries in their respective languages, could differ purely because of linguistic compulsions and variations in meaning accorded to them. Sheer semantics is not the key to the puzzle and a comparative analysis reaffirms that it is not really so. Variations have penetrated due to differing state practices, diplomatic perceptions, national policies and even vested interests. These reasons have created a medley of delineations, each with a different connotation or bearing a divergent nuance for a variety of reasons. This confusion is not to be acceptable in law, international relations and state practice. It is, thus, necessary to arrive at an authentic definition which could be accepted universally in the glossary of space jurisprudence.

The Russian Terminology

The Russian equivalent and equally authentic terminology used in space law and literature and particularly favoured by the Soviet scientists and legal luminaries is "cosmonaut" or *kosmonavt*. Apart from main differences of linguistic character, it departs in its connotation and encompasses a broader scope with a futuristic tinge in its reference to the cosmos. To quote from the *Soviet Space Encyclopedia*, "*Kosmonavtika*", a cosmonaut is a person, who is specially trained in medical, biological, scientific and technical fields and who has participated in a space flight as a pilot-commander or as a member of the crew.[7]

The emphasis in this definition is on training of the cosmonaut and his role aboard the space ship. Here, the purpose and qualifications for inclusion in the crew are not clearly stated and the composition of its members appears flexible. Incidentally, it is also rumoured that Yuri Gagarin in the Soyuz was a subject of research and flew in a standing posture. He was not the crew or commander of a space vehicle which was automated to perform its functions. Moreover, exclusivity of definition is not properly delineated due to the vagueness of the phrase 'as member of the crew.'

Further, a generic impression is that the Communists are atheists and do not believe in God. Atheism has always been a defining marker of a Communist and an inalienable part of Marxism. This coaxes us to premise that Communist states have no ostensible religious affiliations. These states neither patronise nor profess any religion and, accordingly, there is no religious bar or preference whatsoever for eligibility of crew. The Communists are, thus, steeped in the materialistic view of the cosmos and trust only empirically verified physical facts. If at all, Marxism is their religion and true believers expect to be rewarded "with paradise on this side of the grave."[8] Gradually, over time, space technology has become a fetish and surrogate for religion, particularly for the cosmonauts. In corollary, religion accords no preference or seniority to the cosmonauts as envoys of mankind.

It is, therefore, natural to expect that the Communists generally do not accept Christian beliefs about the creation of the universe and man. This is reflected in the fact that after two successful launches of Sputniks, a Soviet statement, rather irreverently claimed that man had invaded the heavens and God was nowhere to be found. There is no paradise and it is pitch dark up there.[9] Emboldened, they did not care to factor in God's will in their safety equations for space travel. Again, in 1960, during the Lunik probes, Radio Moscow proclaimed in a blasphemous tone that "our rocket has by-passed the moon. It is nearing the sun, and we have not discovered God.... We are breaking the yoke of the Gospel....and Christ shall be relegated to mythology."[10]

However, notwithstanding such pronouncements and later the fall of Soviet Communism, the state relaxed stringency over individual freedom to hold religious beliefs and undertake practice of its rituals in private with the revival of the Orthodox Church. This appears agreeable to the inner feelings of the cosmonauts so inclined. In fact, many of them have carried pocket-sized personal charms and amulets of Virgin Mary or Christ on their voyages. Earlier, Mir, and now the Russian portion of the International Space Station (ISS), has photos of churches and shrines, apart from Christ and Mother Mary. Hence, the Soviets are also now flaunting religiosity and spiritualism.

Further, it may seem paradoxical yet the Orthodox Christian Church is supportive of humanned space flights and has been invoked by Russian space agencies to depute the clergy to anoint its rockets and satellites at pre-launch ceremonies for success of the mission, before blast-off from Baikanour. Even cosmonauts are blessed by "holding a golden cross and sprinkling them with holy water."[11] Interestingly, Yuri Gagarin, the first cosmonaut, has been canonised as a saint and patron of space travel and is propitiated by rituals as *"imitatio dei."* April 12, the day of his ascension to outer space, is annually observed as Cosmonauts Day.[12]

The Chinese Nomenclature

The Chinese, too, have their space vocabulary in their own specific language. It is, of course, understandable that their terminology and procedures would have mainly been derived from Soviet manuals and lexicon yet the linguistic differences could be stark. The astronauts in Chinese parlance are called *taikonauts*. This word seems to have been derived from the Chinese language, *taikong* meaning space, and combined with the Greek word *naut* meaning a sailor. Of course, there is an alternative Chinese term, *yuhangyan* that means sailor in the universe. However, no proper characterisation of such a persona is available except for its direct relevance to space flights.

For the Chinese, space is solemn and awesome, hence, there is no doubt that they have rather stringent selection procedures for *taikonauts* that are multi-processed and multi-layered.[13] One of the preliminary criteria for screening and elimination from selection is offensive body odour, which is considered an important health factor and a prime concern of ergonomics for work comfort in the narrow confines of a space ship. Therefore, candidates with foul breath or discomforting body smell or irritating sweat are rejected at the initial screening stage itself. This emphatic concern for olfactory sensitivities deserves to be commended and emulated.

Further, no cut of any type anywhere on the body, even of caesarian delivery, is acceptable for fear of opening up in a weightless environment of space. For women *taikonauts,* those with a successful pregnancy and normal delivery are preferred. Liu Yang, the first Chinese female *taikonaut,* who launched into space on June 16, 2012, in the Shenzou-9 capsule for rendezvous with the Tiangong-1 space station module[14], qualified on this parameter. And, interestingly enough, the final clearance of a prospective *taikonaut* comes not from his own volition or express desire but the will and approval of the spouse who seems to have the ultimate say in endorsing his/her sojourn to outer space. Perhaps a little undemocratic, but it may add to durable conjugal happiness.

The Indian Term

Incidentally, India has devised its own term to address an astronaut and this is *wyomanaut*. The *wyoma* part of the word has its roots in the Sanskrit language of the old Vedic scriptures. *Wyoma* stands for space or sky. *Naut* is a Greek word that means a sailor. Having considered several alternative terms, this word has only recently been finalised by the Indian Space Research Organisation for its humanned space flights for low-earth orbit to be launched in 2016. For this programme, India would soon embark on the process of selection and preparation of *wyomanauts*. It is to be expected that the conditions and criteria for selection would be the same or similar to those of other such space-faring countries.

However, India being a multi-ethnic and multi-religion country, there could be some difficulties in selecting a suitable candidate from certain communities for reasons of their religious injunctions.[15] Take, for example, the Sikhs: the devout are expected to wear, at all times, the five symbols of religion.[16] Some of these like the *kara* and *kangha* may interfere with magnetic components or as loose headware under the spacesuit. Sikhs are also enjoined to wear a turban in all public appearances and may be reluctant to wear space helmets.[17] The religious mandate is strict and violation attracts religious sanction (*tankha*) from the clergy. It remains moot whether a prospective *wyomanaut* will get any circumstantial concession from religious observance. Some sort of compromises may be evolved out of actual necessity.

Another example could be of Muslims. They are ordained to pray at least five times during the day and observe the concomitant ritual of *wuzzu* requiring washing of the face, hands and feet before the prayer (*namaaz*). Again, the Muslim prayer requires the devout to assume different postures of standing, kneeling and sitting down at different times during the prayer. This routine may be difficult to be observed repeatedly in the gravity free environment of outer space. Further, Muslims pray facing Mecca (*Qibla*) which may be difficult to determine or adjust or may even pass into a different direction during the prayer due to the fast speed of the space vehicle. Similarly, Muslim women need to wear a *burqa* or *hijab* in front of persons other than the husband or male blood relations or other very close relatives. The problem comes in focus and seeks due condonation. This also highlights an important aspect of sensitivity to cultural and religious beliefs of fellow astronauts.

The time of the prayer may also come in conflict with real time because the space objects observe the virtual time of the country of launch or jurisdiction, therefore, sunrise, sunset and other prayer times at both places may occur at variance. The confusion, of course, is obvious but a devout Malaysian Muslim, Dr. Sheikh Muszaphar Shukor, has already been in outer space and has observed the rituals religiously. He prayed facing the earth in general but the possibility of the earth being at a position below the spaceship at that time cannot be ruled out. In another similar case, that of a Sultan of Saudi Arabia, he was granted a concessionaire waiver by the *mullah* from regular prayers while in the environs of outer space.

There is, however, a general relaxation in Islam that a person under restrictive health conditions may offer prayers (*salaah*) in any convenient position e.g. sitting, seated or even lying on bed. This practice is accepted among patients, pregnant women and old persons. The indulgence may be logically extended to the astronauts due to the conditions in outer space, but religion is a matter of faith and not logic. Solutions may be individual but the difficulty remains, nevertheless.[18] Religions may have other objections too, even to the forays into outer space because the sky and beyond is believed to be the House of God and intrusion there could constitute an impious act that may render the person pagan.

The US Definition
The US definition states that astronauts are, firstly, those persons who have already accomplished a space flight and, secondly, the test-pilots chosen for participation in different space flight projects.[19] This definition is too pervasive and open-ended. It also includes the stand-by test-pilots, intended and trained for space flights, who, for some reason, did not actually make it to the space-launch as astronauts. On this basis, an astronaut can be one who has not even orbited in space or experienced a blast-off in a spacecraft or is still a trainee undergoing training at an authorised base. Perhaps, this looseness in definition has been intentionally kept for insurance cover and liability factors in unforeseen contingencies.

Moreover, the US definition takes no cognisance of the purpose or role or activity of such a person onboard the spaceship. Hence, as per this definition, even passengers travelling in the space-vehicle would meet the criterion to be called astronauts. This is apparently anomalous for valid reasons. It, therefore, appears that in the American view, the *sine qua non* of an astronaut is selection and training for a space flight rather than actual travel in space. This definition is certainly open to criticism for its unwarranted broadness and loose inclusiveness.

However, the National Aeronautical Space Agency (NASA) has its own protocol for differentiation and classification. The astronauts under training get a Silver Pin and after a successful training flight, are presented a Gold Pin. And the lucky ones who actually fly above 50 miles or 80 km are awarded Astronaut Wings and these get the unique privilege to join the distinguished Astronaut Corps.

The Robonauts
Another confusing term of recent origin is robonaut. Though robots in some form have been in existence for nearly a century, yet with greater

exploitation of Artificial Intelligence (AI) techniques, robots have become humanoids and have further attained dexterity in the use of both hands and are attached with leg-like contraptions for movement on rugged surfaces and other mobility control devices for extended finger and thumb reach. And these are relatively cheap to produce and operate. No wonder, robots are replacing humans in space missions but it must be accepted that robots are mere mechanised and programmed objects with involuntary actions that may not always be innocent or harmless.[20]

The US has elevated them to robonauts and dispatched one of them to the International Space Station (ISS) on the Discovery shuttle flight on February 24, 2011. The humanoid robonaut R-2[21] is addressed as space crew and is expected to work for the benefit of, and alongside, the US astronauts in the Destiny lab though it would be remotely operated from the ground control for the present. These have since become the symbol and the mascot with an image that is striking and apposite. However much the robots may be touted as robonauts, being efficient and less venal, these have inherent limitations of artificial intelligence and they lack human perspective; and capacity to react to emotions and adjust according to situations to get the best results under all situations and in variable circumstances.[22]

Be that as it may, in fabrication robonauts are anthropomorphic mechanical machines that can act as electronic surrogates of astronauts and can be deployed to explore hazardous reaches and even colonise outer space. They can be tasked on autonomous repair operations as they are capable of handling Extra-Vehicular Activity (EVA) tools and interfaces. They can also be usefully deployed for dangerous jobs outside the space-ship as also for long duration or risky deep space probes. With increasing versatility and mobility, they are sure to expand their territory and applications in due course. In fact, such a development had been foreseen and anticipated for some time.[23] As a result, R-2 is a singularly pioneer robot that is state-of-the-art in technology as well as humanoid in cuteness and charisma. It is expected to undertake tough and stoic tasks in a relentless quest and has already attracted hyped media attention that may pale the glory of true-blood astronauts.[24]

Robonauts offer great advantages in work situations, hazardous tasks and in the ultimate cost-effectiveness. These are economical in fabrication and programmed training as also frugal in resource consumption, for example, they need little maintenance and practically nil support for sustenance in food and water. This has, in fact, been a major challenge to balance compulsive food weight and resource utilisation.[25] Robonauts can survive without the controlled, narrow range of temperatures required for human astronauts. However, their true utilitarian economics rests in the fact that these robots are expendable, dispensable, disposable *in situ* and need not be returned to the earth. It would certainly make for considerable saving on mission expenses. They are available at mere $ 2.5 million apiece.

Varying Terminology in Space Law

A comparative analysis of the existing corpus of space law comprising treaties, conventions, agreements, guidelines, principles and domestic legislation reveals an extensive use of the term astronaut which has referred to different categories of experts in different situations. It has been used variously to indicate varied types of space-farers who have, in several texts and contexts, been alluded to, and addressed, differently. Sometimes, even connotations or nuances have varied significantly. A few examples will vindicate this viewpoint.

Even a cursory reading of the germane law throws up a wide variety of terms that have been used to allude to the genre of astronauts. The varying nomenclature may carry different understanding in popular parlance, yet has been applied to refer to the same persona from different perspectives. It should, therefore, be least surprising to find that even the OST, being the *grundnorm* of space law, mentions variable terminology and its usage relating to the word astronaut is not at all consistent or uniform. Thus, confusion is endemic and contagious to similar instruments.

For instance, the term astronaut has been used in Article V of the Outer Space Treaty but the same personae have been referred to as "personnel" in Article VIII, while the Russian text of the same Article of the OST uses the term "crew" and it is "representative" in Article XII. The apparent variation shows a differing treatment or viewpoint. Similarly, the Rescue of Astronauts Agreement,[26] in its Preamble, mentions the term astronauts, while in Articles 1-4 of the same agreement, the words used are "personnel of a spacecraft". Interestingly, the French text of the agreement refers to personnel as *"equipages"*. Similarly, the Convention on Liability,[27] in Articles 3 and 4 uses the phrase "persons on board a space object".

Thus, different terminology has referred to the same persona in different contexts and from different angles because no common acceptable definition of an astronaut has been enshrined in any treaty, convention, agreement or lexicon. This is, indeed, lamentable, even conceding that negotiating an acceptable definition is no easy task at conventions. Positive convergence on this issue is urgently called for to eschew any ambiguity whatsoever.

A Proposed Definition

The necessity and importance of a proper definition of an astronaut cannot be wished away by default and it would perforce have to be evolved through legal debate and consensus. Our concern here would be to arrive at a definition of an astronaut with a legal personality which is singularly differentiable on diplomatic protocol, tenable on precepts of international law and acceptable in international polity. To create such a *persona juris*, its salient attributes will have to be culled out from various legal provisions contained in legal lexicons, treaties, conventions and agreements.

The Criteria

Based on legal analysis, protocol mandates and state practice, the following criteria can be suggested:

Selection and Training
- That the astronaut should have been so selected through a proper procedure, imparted the necessary knowledge, trained in the assigned job and possessing skills as a professional to perform or execute a duty or task or a specific mission on board the space vehicle.

Professional Duty
- That the activity of such a person should encompass any function or role or duty or profession like pilot, navigator, engineer, doctor, physician, biologist, chemist, technologist, observer or any other kind of specialist involved with the instant space mission or even as a subject, medical or otherwise,[28] for experimentation or observation during a space flight.

Lawful Activity
- That the activity of such a person should be connected in some way with the legitimate scientific exploration or peaceful use of outer space or lawful activities on the Moon or any other celestial body to be undertaken in accordance with the principles and rules of international law and other relevant laws. Astronauts undertaking clandestine or unlawful activity, if and when known, shall not be designated, or will be stripped of, the status of envoys of mankind.

On Board a Space Ship
- That the activity of such a person should, or be planned to, take place on board the space vehicle or outside of it while in outer space or on the moon or another celestial body.

Contingent Status
- That a person who satisfies the above criteria, but perchance due to some mishap or accident, the vehicle in which he had to make the flight, returns without entering outer space or makes an emergency landing or crashes or disintegrates or is destroyed in any manner after commencement of blast-off, would still qualify under the above criteria despite being the crew of an aborted mission and would be deemed as an astronaut.

An Amplification

It may be amplified further, that despite the above attributes of an astronaut, the following categories of persons connected with and/or involved in

space activities will be excluded from claiming to belong to the corps of astronauts:
- A person selected and trained for space flight but who does not actually and physically travel in the space ship on a duty related to the space mission. The US definition appears rather slack on this aspect to include such persons within the ambit.
- Passengers who use a space ship merely as a mode of transportation whether for the thrill of a joy-ride or as spare crew/technician/scientist to reach their place of duty or activity on the moon or other celestial body or a space station or another orbiting space ship, would not qualify as astronauts during such space flight. This is because they have no task to perform on board this flight. The suggested strictness is introduced in the context of being designated as an envoy.
- Specialists or technicians who work inside the space vehicle for its ground maintenance and preparation towards blast-off or operate infrastructure outside the space ship, and are to remain only earthbound, would not attract the status of an astronaut.

A Proposed Definition

We may now venture to evolve a suitable definition. An astronaut is a person who is a professional so selected and trained and assigned a duty on board a space ship during or after its blast-off or on any space station or the Moon or any celestial body towards achievement of the mission of that particular space ship and whose activities are connected with the exploration and use of outer space in accordance with the principles and rules of international space law, provided, of course, that the space ship had commenced the blast-off.

It is readily conceded that the definition expounded above is descriptive and may fall short of strict techno-legal mandates. But it can certainly serve as a good starting point to elicit refinements or amendments. It affords a tentative premise for international space law which can be debated, appropriately modified and, if necessary, redefined suitably for embodiment in any international treaty or convention or lexicon on germane subjects for universal acceptance.

Clarifications in *Abundanti Cautela*

Passengers Are Not Astronauts
As discussed earlier, the US and Soviet definitions of astronaut have been found to be loose, vague and grossly inclusive. According to their stipulated parameters, space tourists could easily qualify for the status of astronaut and, by inference, envoy of mankind in outer space. But such a supra-national position of privilege to every tourist is legally paradoxical and fraught with intrinsic contradictions. This contingency would add to the problem of too many at the same time as well as the clout of lucre to secure such coveted representation.

Further, from the proposed definition, it logically follows, as a corollary, that fare-paying passengers and space tourists cannot be classified as astronauts. In fact, it makes no business sense, nor are the official obligations of an envoy any good for them because they may interfere with their pleasure and excitement during the mission and, at the same time, it will be well nigh hard to determine their precedence for protocol. Even the Russian expedient of calling them "space-flight participant" seems an equally inappropriate differentiation between a paid-traveller and true astronaut.

On the contrary, even the fare-paying tourists have resented this status of an astronaut and the attendant implied responsibility, despite assertions of mission-activities in outer space. Dennis Tito and Shuttleworth claimed to have conducted vital experiments on board the Soyuz flight as well as in the International Space Station (ISS).[29] They had even signed the Code of Conduct incumbent upon participants and residents at the ISS, yet they are believed to have mostly tended to focus on their unique, exhilarating space experience that they had actually paid for. Some other space tourists also have been vocal on this issue to eschew this categorisation.[30]

Personnel on Unlawful Missions Not To Be Envoys
It seems pertinent to suggest that an astronaut so assigned and performing a patently unlawful activity or undertaking an experiment prejudicial to the peace and security on earth or in any manner causing injury or damage to the environment of outer space, intentionally and advertantly, should not knowingly

be designated as an envoy of mankind and if already so appointed by a competent authority, should be stripped of any such status and accreditation nullified to be treated as *persona non grata*, though in the limited sense.

It is hoped that the above proposed provision would act as a deterrent to illegal and clandestine activities contrary to the best interests of peace and security on the earth. Such an action to disqualify would adversely affect the prestige and standing of the parent country and would be considered least flattering or desirable. It is to be expected that many jurists would look askance at such a suggestion and may even vehemently argue that no international authority, not even the UN, is competent or equipped for such a declaration. If it is believed that such a provision would better the interest of mankind, then ways and means to implement this can be found or evolved. The stakes are ours — so is our ingenuity to fend for this.

Astronauts as Envoys of Mankind in Outer Space

Astronauts as Envoys

International law and diplomatic practice[31] recognise accreditation of an envoy as an emissary to a receiving state for the protection of the interests and activities for the benefit of the sending state or for negotiations through intercession of professionals with a non-confrontational stance, to find mutually acceptable solutions to a common challenge. For this purpose, the envoy as the representative of the sovereign government or the executive head of the state, enjoys certain privileges and immunities in the territory/ies of accreditation.[32] Their powers are sized according to their diplomatic status, whether ordinary, extraordinary or plenipotentiary.

Recalling this customary practice of states from ancient times,[33] the Outer Space Treaty declares that astronauts shall be regarded as envoys of mankind in outer space. This certainly is a coveted status accorded to the astronauts, prompted by the belief that the activities of the astronauts in scientific exploration and other peaceful uses of outer space shall be for the benefit of mankind as a whole. Thus, great significance has been attached to their mission in furtherance of human knowledge and the welfare of the peoples of all nations. It was fervently hoped that the results of space probes and facts gathered by missions into outer space shall be harnessed and used towards the common interest of humanity. The intentions were noble and the expectations quite legitimate.

Envoys of Mankind

The Outer Space Treaty was the first multilateral agreement to use the term "astronaut" but failed to ascribe to it a proper description of its profile. And again, it was this treaty which declared astronauts as envoys of mankind. The relevant portion of the text of Article V of the treaty mentions,

> States-Parties to the treaty shall regard astronauts as envoys of mankind in outer space and shall render to them all possible assistance in the event of accident, distress, or emergency landing, on the territory of another State-Party or on the high seas. When astronauts make such a landing, they shall be safely and promptly returned to the state of registry of their space vehicle.

The Outer Space Treaty reveals a commendable urge for cooperation and collaboration, and the supra-national status accorded to the astronauts is, indeed, laudable. As envoys of mankind in space, the astronauts would, thus, enjoy an exalted status and distinguished privileges guaranteed for diplomats under international law, state practice and the customs of civilised nations. The treaty enjoins state-parties to accord them special immunity irrespective of their actions or circumstances of unprogrammed landing and "tender all possible assistance" to ensure their safety and prompt return to the state of registry of their space vehicle.[34] This contention is independent of the provision pertaining to jurisdiction over objects and personnel under the Rescue Agreement.[35]

This provision, indeed, imbues astronauts with a very distinctive character and an exalted status as representatives of mankind in outer space and, in turn, casts on the states an *obligatio juris* under international law and the customary practices of states. Perhaps, they were intended to be messengers of the mother earth to other planets, should they discover or encounter any like-form of life in the cosmos. The treaty had anticipated and catered to such a contingent situation.

The concept was, of course, commendable and exhibited an excellent façade of unity and integrity among the peoples of the world. State-parties to the treaty had, perhaps unwittingly, agreed to a utopia; possibly, under the euphoria of the technological successes in outer space. Moreover, it is interesting to note that the term "astronaut" has nowhere else, except the OST, been alluded to, or exemplified, as an envoy of mankind. Even subsequently concluded treaties do not use this phrase. Not even the Rescue of Astronauts Agreement mentioned above. On the contrary, this agreement appears to be at pains to avoid even the use of the word "astronauts" and all along refers to them, in its text, as "the personnel of a spacecraft",[36] even at the risk of confusing between the true astronauts and the fare-paying passengers.

Concept of Mankind[37]

The treaty declaration that astronauts shall be regarded as "envoys of mankind in outer space" compels us to dwell on the concept of mankind. It leads us to an interesting paradox, as to whether mankind is an aggregate of states and a simple summation of the consensus of states constitutes the will of mankind. Or that mankind is more than a mere conglomeration of states and is a separate entity of discernible traits with an independent existence and character. And can mankind be a subject of international law and does mankind have a legal personality and independent volition with capacity, *sui juris*, to act in its own name? Or would it be entitled to be the bearer of its own rights and the trustee of the interests of all those it seeks to comprise, represent and promote? The space law literature is, predictably, silent on the above issue. But one may rationally deduce the following chain of logic to arrive at a coherent synthesis and an understandable concept.

Mankind as a whole can arguably be perceived as a discrete and distinct entity apart from, and more than, an aggregate of all the states. Mankind can be the beneficiary and bearer of legal rights and interests vested in it, like "common heritage" and "common interest" and the "province of mankind" under various conventions e.g. the Laws of the Sea or Antarctica. But mankind lacks the active legal capacity to exercise its rights directly and protect its interests against subversion or exploitation. It is unable to act on its own under international law and would need a mechanism or instrument in the form of an international organisation like the UN to act for, and, on, its behalf. And in the ultimate, the common benefits and fruits reserved for mankind are to be actually received and ideally shared in equitable proportion by the states which compose and constitute mankind.

Thus, mankind apart from being a conglomerate of states, has an independent and distinguishable existence. It certainly has a limited personality in law but without active and inherent capacity, under the captaincy of a competent institution, to enforce its rights. In a nutshell, it has a passive legal personality. Now we may introduce another parameter to the analysis. Mankind is not only a spatial or existential concept, it has a futuristic dimension also. The true notion of mankind includes future generations as well and the insurmountable poser is: who can legitimately represent the rights and interests of posterity? The *locus standi* is debatable.

Political theorists have glossed that every state has the *locus standi* to represent its future nationals in connection with *erga omnes* obligations.[38] But it is regrettable that "the States being political units perceive the future as divided among States and find it hard to liberate themselves from the domestic constituencies [and this] prevents them from promoting the cause of mankind as a whole. Thus, they perpetuate the divided world instead."[39] Under this mindset and vested compulsions, the states will never be in a position to assume an objective and impartial stance essential for reaching a just and fair decision in a conflict of interests between a nation and mankind. The living reality stoically overrides the future premises. Mankind, therefore, needs either an international organisation as advocated by N.M. Matte[40] or an entity in the form of an ombudsman as suggested by E.B. Weiss[41] that can impartially and objectively cogitate and equitably decide on issues of common heritage and allied matters in conflictive situations of sustainability between the existing populace and future generations, collectively called mankind.

Only in Outer Space
It is also pertinent to mention that astronauts shall have the right to bear and uphold the status of envoy of mankind only so long as they are in outer space or on a celestial body. Because this diplomatic status ceases

once they return to the planet earth and assume their original nationality to relinquish this haloed personality. Frankly, it is not even necessary for the astronauts to carry this notional designation on earth because here they may neither encounter aliens in a similar circumstance nor would they be responsible to put up behaviour conforming to the protocol and etiquette of an envoy. Undoubtedly, other mechanisms and modalities exist on the earth to effectively handle such eventualities.

A Legal Appraisal
Before embarking on an analysis of the legal basis, it would appear pertinent to discuss the historical sequence and make a reference to the material contained in various committee reports to enlighten on the purpose and intent of the provision phrased as "envoys of mankind" at the time of its inclusion. It now appears that since then, technology has taken grand strides in the space domain, while many new international legal instruments have been negotiated that have significantly altered the import of this treaty and the attendant scenario. No wonder, the clause has lost most of its original intent, notional standing and practical relevance.[42]

The UN General Assembly in 1958 had instructed an Ad Hoc Committee to report on "the nature of legal problems which may arise in the carrying out of programmes to explore outer space".[43] A subsequent resolution directing the Committee on Peaceful Uses of Outer Space (COPUOS) forcefully reiterated the above general mandate.[44] It was only in 1961 that the General Assembly touched tangentially on a factor of sensitive practical relationship *albeit* involving humanitarian considerations relating to rescue operations and possible succour to astronauts and return of space objects in the event of accident, distress, emergency or unintended landing. In the absence of a suitable provision in any convention, there was an urgent need to arrive at the substantive core of a proposed understanding in a general agreement and an enunciation of the basic principles for universal acceptance and adherence. Cooperation was the fundamental preaching as well as the need of the hour.

Later, in 1962, the legal sub-committee of COPUOS was given a clear mandate to broadly formulate a legal regime for international cooperation in the outer space environment.⁴⁵ This culminated in the "Declaration of Legal Principles Governing the Activities of States in the Exploration and Use of Outer Space" that was adopted by the UN General Assembly.⁴⁶ The declaration prescribed, *inter alia*, that "States shall regard astronauts as envoys of mankind in outer space..."⁴⁷ Later, the same clause was incorporated in Article V of the Outer Space Treaty, 1967, for an incumbent state practice and to elicit customary cooperation from the states.

It becomes evident from the above that the members of the General Assembly honestly intended to accord the astronauts an exalted status, thereby providing international protection and inviolable privileges. It was realised and recognised that the profession of astronauts involved ultra-hazardous actions and unknown imponderables. Hence, the international community owed a solemn responsibility towards the security and safety of the astronauts and was, thus, enjoined upon to cooperate for their personal safety, prompt rescue and legitimate return.

It is interesting to note that the phrase "astronauts be regarded as envoys of mankind" has not been formally used elsewhere after the Space Treaty of 1967. In fact, it appears that states became reticent even to use the term 'astronauts' after this treaty. Agreements formalised after 1967 have variously described astronauts as crew or "their personnel"⁴⁸ *et al.* Even the Agreement on the Rescue and Return of Astronauts and the Return of Space Objects desists from conveying any impression that the astronauts are envoys of mankind and merely refers to them as "personnel of a spacecraft". The motivation for the *volte face* appears evident and needs no elaboration.

When the phrase, "envoys of' mankind in outer space" is considered with reference to the context in the UN Declaration of 1963 or the Treaty of 1967, it transpires that the clause was used in relation to the need to ensure the personal protection and safety of limb and life of the crew, so essential in the solitary and isolated environment of outer space. It was also possibly prompted to elicit international assistance for their security and safe return to the launching state. Therefore, the use of the phrase, in the initial stages during the era of the Cold War, had an apparent motive and the vested interest of states undertaking manned space probes, particularly before the Rescue of Astronauts Agreement had been concluded.

Even legally, it created an interesting paradox where the international community, though subservient to the concept of territorial jurisdiction on the earth, under the established international law, could be bound by *specialibus* treaty law pertaining to outer space to "promptly return the crew to the launching state" without let or demur. This rule also applied to objects in defiance of customary jurisdiction. Further, this mandate also applied in the eventuality of a rescue from a celestial body or outer space where and while they retained the status of envoy of mankind at large and concurrently owed unabated allegiance to the state of registry or launch. This problem has become magnified and complicated in space activities with multiple-numbered and multinational crew.[49] Myriad implications can be surmised.

The Dilemma and its Resolution

The instant provision in the Outer Space Treaty is laudable and forward looking, but appears badly phrased for several reasons, if not intentionally so. This creates a strange dilemma of paradoxical situations, bordering on the comical. The problems pertain to concurrent multiplicity of envoys, determination of *inter se* seniority of multinational crews, issue of identity documents and accreditation by an apex authority. Apart from these problems, the situation is endemic with avoidable confusion, imaginable disputes, foul diplomatic etiquette, and human proneness to misbehaviour. These points are discussed in the succeeding paragraphs.

The Dilemma

Science and technology have taken great strides where the size of space ship structures is concerned.[50] As a result, the crew of a space ship is generally more than a solo pilot and, in most cases, comprises multi-numbered crew to ease up on space traffic and for economics of costing. Apart from specific crew, the space ship may also carry additional personnel on board as experts or subjects who too may technically qualify for the status of astronaut under existing definitions and *ad idem* understanding. In such a situation, a spaceship may have several astronauts and, should the need arise to act as envoy of mankind of the earth, there would be a multiplicity of such claimants. Among them, who would be considered senior or superior to take precedence over others is not laid down in the treaty or any subaltern rules or principles. Though silent on this aspect, yet surely, the intention of the General Assembly Resolution[51] would not have been to treat all the astronauts horizontally and simultaneously at any point of time on a particular space ship, station or celestial body to act together as multiple or competing envoys. Diplomatic courtesy requires a vertical selection and designation of one among them to represent, whatever be the criteria.

In today's reality and state-of-the-art space shuttles, space transportation and space tourism are just round the corner. With such a scenario, besides multiple crews, multiple-passenger capsules for space tourists would be common practice for convenience in operations and economy in fares. Under the existing definition, the tourist also acquires the status of an envoy and, thus, multiplicity of envoys can create not merely an anomalous situation but a near perilous diplomatic order that will certainly not reflect well on the superior intellect of humankind. Hence, passengers and tourists cannot be deemed as astronauts and symbolically dignified as envoys of mankind.

Next, it is not an unlikely eventuality that the first contact with sentient aliens may occur with a robonaut on a deep-probe space mission. A 'humanoid astronaut' will, thus, stand for a 'human astronaut.' The contact and representation would be utterly unrealistic. And in such an encounter, it can well be surmised as to what impression aliens take of the looks and intelligence of *homo-sapiens* inhabiting the planet earth. The situation seems least flattering and infuses no pride.

Cooperation is the key word in space law and space activities and in deference to this principle, space-faring nations have started exchanging personnel to participate reciprocally in scientific explorations and peaceful exploitation of outer space and the celestial bodies.[52] The International Space Station (ISS) is an outstanding example of such collaboration where different countries have their research modules and astronauts work jointly and severally, depending upon the nature of the experiment or operation.[53] This has led to the situation of multinationality crew operating within the same precincts and ambience, in immediate contact and a enclosed environment. This situation again creates a multiplicity of astronauts *alias* envoys at the same time and place It makes no sense to let them behave on instinct or intuition in dire contingencies. There can be only one true representative but what are criteria to designate such a one? Some options could be seniority of service, length of experience, or superiority of nationality in terms of political or financial standing.[54] The dilemma is sharp and clamours for a just resolution.

It has been customary state practice that envoys have to be selected, trained as diplomats, briefed and accredited to act as agents with credentials of a sovereign authority so as to be able to be the formal and exclusive representatives.[55] The same practice and methodology should apply to the envoys of mankind. But if this be so, then who undertakes and controls these preliminaries and diplomatic authorisations? It cannot be mankind as a whole being a non-discrete, fluid entity; then, would it be the UN or Office of Space Affairs (OOSA) or any other designated authority or the launching state?

It seems interesting to mull on the issue of whether other states can by virtue of being part of mankind, influence, monitor or control the processes concerning astronauts who are to represent them as part of a collective whole in outer space. And if these other states are helpless in this regard, then would they like to willingly permit these astronauts to be their envoys too? And can a state or group of states resent or protest the representation collectively or severally or withdraw concurrence for any particular astronaut to act as an envoy on its/their behalf as part of the world comity? The dilemma is clear and stark and forebodes, at the least, serious political embarrassment.

Thus, a possibility could be visualised where states with no stake in, or contribution to, the exploration of outer space could hold space powers accountable or at least discomfit them for the acts of omission or commission by their astronauts. Such a disparaging situation had to be avoided in the future. As a result, the space-faring states surely did not wish to repeat the mistake of continuing to regard astronauts as envoys of mankind and supplanted this dictum with the principle of international cooperation and concurrent assertion of national jurisdiction over the crew and the space objects. And it has since been the nationalistic chauvinism of space powers that has prevailed for the last half a century and has come to be thrust as an obligatory maxim of space law.

It seems quite likely that states, at the stage of drafting the UN Resolution, had not fully realised and appreciated the true import and implications of such a declaration bestowing the status of envoy of mankind on each and every astronaut, despite the professed innocence and magnanimity. It seems to have later on become apparent to the space powers that it may not be unthinkable for the majority of small, undeveloped, non-space-faring states, by the sheer arithmetic of numbers in the General Assembly, to start clamouring for proportional right of control over the astronauts or insist on their subservience to the concepts of "the benefit of all peoples" and "the common interest of mankind " so loftily enshrined in the OST.

The Resolution
Since the days of the OST, there has been a lot of change of thought due to altered political equations of the Cold War and subsequent cooperation elicited by later treaties and agreements. There was a realisation that this dictum of astronauts as envoys has outlived its utility. With multinational crews onboard, the situation got further dented by the Convention on International Liability for Damage Caused by Space Objects, 1972, which does not apply to foreign nationals during such time as they are participating in the operation of that space object or instant mission.[56] Space tourists are not expected to fall within this exception and it may turn detrimental to their interests. Therefore, the tourists, in a technical sense, cannot be admitted to the *corps de astronaut* and should not be burdened with additional responsibility of an envoy of mankind in outer space.

In general, the predicament is endemic with confusion, disputes, conflicts or even fights. Taking a cue from an analogous medium or environment, it would seem prudent that only the commander of the spaceship should be designated as 'the envoy' for that particular mission while in outer space and not all the astronauts onboard as a generic category of envoy. Similarly, on the moon and other celestial bodies also, some sort of hierarchy or rotational privilege would have to be contemplated and implemented for legal order and stable governance so as to obviate any conflictive squabble or unsavoury situation.

Apart from accreditation, it appears imperative that the envoy should carry a document of identity, comparable to a national passport, for being an "Inhabitant of Planet Earth" (IPE) or "Member of Humankind"(MoH), to whom he or she should belong and would be representing. It can also be mooted that we graduate from mere citizens to a higher echelon of "cosmozens" possessing an eclectic mix of cherished human plus cosmic values. Apart from this identification, some kind of credentials may also be issued as authorisation to officiate in such designation. Here again, the UN could issue such documents when requested by the mission-sponsoring state. This may not be the best resolution of the tricky problem, but certainly constitutes a benchmark for further refinement of the proposal.

Conclusion

Intoxicated by the euphoria of the conquest of outer space as a human accomplishment of global pride, states tended to honour astronauts for their daring feats of adventure. At the same time, the space powers were also conscious of possible eventualities that accidents could occur, emergencies could befall and space crew could be in distress, needing help or assistance while on celestial bodies, in outer space or on territories of unintended landing or crash. Consequently, realising the need for assured reciprocal cooperation and worldwide coordination in rescue operations and handling of emergencies, the space powers found it convenient and expedient to characterise astronauts as envoys of mankind in outer space.

This mandatory accord of an exalted status was intended to impose an obligation on all states to extract all possible help and assistance, when in need, for its space heroes, on professedly humanitarian considerations and rather slender diplomatic nuances. Perhaps, it was a compulsion of the Cold War era and a ruse to elicit obligated cooperation in space emergencies. Nevertheless, another ostensible object of bestowing this status was that should the astronauts come in contact with sentient life in the cosmos, they would have the due authorisation to communicate and interact.

Howsoever meaningful and relevant, the legality of this provision, from the very beginning, was under a shadow of doubt for various reasons. The controversial points in the procedural formalities related to the issues of diplomatic warrant and accreditation, the nebulous identity of the sovereign authority it sought to represent and the legal status of the collective of mankind as a juridical entity. These dilemmas, but along with other putative problems, remain vexing and outstanding though were soon relegated to blissful oblivion due to obvious embarrassment.

Soon, the supra-national status granted to the astronauts became inconvenient and unacceptable in practice. As a consequence, the space powers quickly usurped jurisdiction over the crew and space objects; eliciting support from the customary legal tenets of "recognition of the flag" and "floating/flying territory" from the Admiralty Law and the Chicago Convention of the aviation regime, respectively. The diplomatic status of astronauts as envoys of mankind soon went into desuetude and nearly faded into legal history. Astronauts were again reinstated as nationals first and foremost before anything else. Therefore, it is now established law that assures national jurisdiction over astronauts and satellite hardware

Apart from the above, the predicament got compounded, when, with advanced space technology facilitating bigger space vehicles and space stations, the concept encountered practical difficulties. First, the definition of astronaut is at variance in different space-faring countries. Second, with multiple crew on board and each one qualified as an astronaut and, thus, eligible to be an envoy of mankind, multiplicity poses serious problem because the law is silent on the criteria for selection. Third, with multiplicity there is the conundrum of the multinationality of the crew, all at the same time, at the same place. Considering seniority as the principal criterion, *inter se* seniority poses ticklish issues endemic of diplomatic discord. The dilemma is obvious and seeks an urgent solution.

To conclude, astronauts were accorded the status of envoys of mankind in outer space but the attendant dignity and privileges lacked sound legal foundations and created an inherently conflictive paradigm. The basic purpose was to elicit cooperation from the international community for the astronauts in times of distress. This provision was, therefore, projected on purely human-centric considerations of utmost concern and was wrapped in the garb of pseudo-legality and core diplomacy. The platitude remained attendant with ambiguities. The espoused generalities and pious intentions did not prove adequate mechanisms to support lofty ideals or for resolution of conflict situations or contentious claims.

Later, when it became obvious that this contention could cause embarrassment to the space powers, they were quick to shun the use of this charismatic diplomatic phrase and the launching states arrogated to themselves "jurisdiction over space objects and their crew". Traditional legal concepts of flag and territoriality came in handy to claim *de jure* control and obtain allegiance from the truant as a customary tenet. International agreements followed to neutralise this innocent mistake and to formalise it into a policy stand. The tenuous legal basis for envoys of mankind had weakened even under the treaty law and this status now remains notional and contingent, should astronauts ever encounter intelligent life in the cosmos.

This provision seems to have outlived its utility and now use of this phrase appears symbolic yet seems to carry the stink of vested interest devoid of any genuine intent of noble design. In actual practice also, this clause appears to have lost its relevance and desirability; and utter lack of its reference in subsequent international instruments makes it rather feeble and almost futile. Nevertheless, it would appear sagacious to bestow this contingent status, and under the given circumstances, to designate only the commander of the space vehicle or of the space station to act as such envoy, *inter alia,* other supernumerary duties for ensuring discipline and proper order on board the space-ship. This option would obviate the possibility of multiplicity of envoys at any given point in time and space.

Again, the envoy must be named individually and accredited by an appropriate authority, preferably by the UN which, however, is not exactly sovereign in the political sense. The envoy must also carry a document of identity and authorisation to represent mankind. Procedures for such a process need to be put in place formally for compliance. The international community would be well advised to act judiciously and fast in this regard to clarify ambiguities through consensus and enunciate a firm stand of common understanding as also evolve the necessary protocol for proper implementation. This is our moment of opportunity before it becomes too late.

Notes

1. Treaty of Principles Governing the Activities of States in the Exploration and Use of Outer Space, including the Moon and Other Celestial Bodies, 1967 (in short the Outer Space Treaty or OST). UNTS, Vol. 610, p. 206 ff. or UN Doc. A/6621, December 17, 1966, pp. 11-18.
2. Ibid., Article V.
3. Discussed later in the Chapter.
4. Agreement Governing Activities of States on the Moon and Other Celestial Bodies, 1979, popularly called the Moon Treaty. Refer Article 10. Words in parentheses have been added.
5. Agreement on the Rescue of Astronauts, the Return of Astronauts and the Return of Objects Launched into Outer Space, 1968. In short, Agreement on Rescue of Astronauts, etc.
6. President Obama, while addressing the astronauts at the ISS and in the Discovery shuttle, referred to the R-2 robot "as new crew member."
7. Kosmonavtika, *Malenkaya Entsiklopediya* (1970), p. 239. Also refer to E. Kamenetskaya, "Cosmonaut: An Attempt of International Legal Definition", *Proceedings of the Thirty-First Colloquium on the Law of Outer Space* (1988), (Publication distributed by American Institute of Aeronautics and Astronautics, Washington D.C. in 1989), pp. 177-178.
8. Joseph A. Schumpeter, *Capitalism, Socialism and Democracy* (New York: Harper Perennial, 1962), p. 5.
9. Virgiliu Pop, "Viewpoint:Space and Religion in Russia," *Astropolitics*, Vol 7, No 2, May-August, 2009, p. 153.
10. Ibid.
11. Ibid., p. 158.
12. Ibid., pp. 155-156.
13. Xinhua, China News Agency Report reproduced in *The Times of India*, March 10, 2009. A selection procedure in 1986 did not yield any candidates. http://taikonaut.asia/dev/index.php?option=com_content&view=category&layout=blog&id=54&Itemid=11
14. Tiangong in Chinese language means Heavenly Place.
15. This seminal idea transpired during discussions with Virgiliu Pop of the Romanian Space Agency at a conference at Hyderabad in June 2012. He has researched extensively on space and religion. The same has been developed and elaborated to make it meaningful and relevant.
16. These are *kesha* (unshorn hair on the head and beard on the face), *kara* (an iron bangle worn on the wrist), *kirpan* (a sword of variable size held on the left side of the waist), *kangha* (generally a wooden comb of suitable size in the hair on the head), *kachha* (typical long underwear reaching close to the knees).

17. In many states of India, Sikhs are exempted from wearing helmets while riding two-wheeler motorbikes. Even Sikh women enjoy this exemption.
18. Learnt from an enlightening discussion with Virgiliu Pop.
19. *Dictionary of Technical Terms for Aerospace Use* (Washington D.C. 1965).
20. The first known case of robot homicide occurred in 1981, when a robotic arm crushed a Japanese factory worker.
21. The R-2 has been developed and manufactured by the Oceaneering Space Systems of Japan that is fast advancing in this direction and hopes to land a robot on the moon by 2015. Accessed from www.fastcompany.com/1653562/japan-robot-moon-base.
22. The robonaut was activated on August 23, 2011. Refer <http://www.thehindu.com/scitech/science/article2389394.ece>
23. Roger Lanius and Howard McCurdy, *Robots in Space: Technology, Evolution and Interplanetary Travel* (USA, 2008).
24. http://www.thehindu.com/sci-tech/science/article_701981.ece?home, accessed on April 26, 2011.
25. As per US standards each astronaut is allocated 3.8 pounds of food per day. Thus, the saving of payload and transportation economics for robonauts would be significant and attractive. For this reason alone, the US is actively involved in research on bio-generative system of growing crops in space e.g. in celestial habitats and on spaceships
26. Agreement on the Rescue of Astronauts, the Return of Astronauts and the Return of Objects Launched into Outer Space, 1968. In short, the Rescue of Astronauts Agreement. It was formalised on April 22, 1968. UNTS Vol. 672, p. 119 ff.
27. Convention on International Liability for Damage Caused by Space Objects, 1972.
28. It is stated in an oblique reference that Yuri Gagarin, the first human being to bear the honour to enter outer space, was not truly commander of the space ship but was sent as a medical subject for verification of the myth of losing sanity on entering outer space.
29. Dennis worked as a scientist at the NASA Jet Propulsion Laboratory for 5 years. He took copious notes during flight and at the ISS and later widely shared his space flight experience. http://wwwspaceref.comnews/viewsr.html?pid. For experiments of Shuttleworth refer http://wwwafricaninspace.com. Both websites accessed on July 27, 2011.
30. Oslen preferred to be called "Private Researcher" and Garriot desired the title "Private Astronaut". NASA has used the term "Spaceflight Participant". Refer http://enwikipedia.org/wiki/Space_tourism#Legality. Accessed on July 26, 2011.
31. The Preamble to the *Vienna Convention on Diplomatic Relations, 1961.*
32. *Vienna Convention on Diplomatic Relations, 1961*
33. In recent times, formalised at Congress of Vienna, 1815 and Congress of Aix-la-Chappelle, 1818.

34. n. 1.
35. n. 5.
36. Ibid., Article 1.
37. The term mankind here means and connotes humankind with no implication of gender bias whatsoever.
38. B. Nagy, "Common Heritage of Mankind: The Status of Future Generations," in the *Proceedings of the Thirty-First Colloquium on the Law of Outer Space*, pp.100-102.
39. Ibid., p. 104ff.
40. Detlev Walter, *Common Security In Outer Space and International Law* (Geneva: UNIDIR, 111, 2005).
41. Ibid., p.272.
42. Refer G. S. Sachdeva, "Astronauts: Envoys of Mankind: An Analysis of Legal Basis," in V. S. Mani, S. Bhatt and V. Balakista Reddy, eds., *Recent Trends in International Space Law and Policy*, (New Delhi: Lancer Books, 1997), pp. 209-217.
43. General Assembly Resolution 1348 (XVIII), December 13, 1958.
44. General Assembly Resolution 1472 (XIV), December 12, 1959.
45. General Assembly Resolution 1802 (XVII), December 14, 1962.
46. General Assembly Resolution 1962 (XVIII). Unanimously adopted on December 13, 1963.
47. Ibid., paragraph 9.
48. Agreement Governing Activities of States on the Moon and Other Celestial Bodies, 1979, popularly called the Moon Treaty. ILM, Vol. 18 (1979), p. 1434ff. Article 12(1).
49. For example, the Apollo-Soyuz link-up mission with the US and Soviet astronauts, or space missions involving Soviet and European cosmonauts, or an Indian astronaut as part of Soviet space ship, or the multinational crew of the International Space Stations or in future celestial settlements.
50. The Soyuz of Yuri Gagarin, with narrow confines, had no manoeuvrable space. The present day ISS is of the size of a football ground. This vindicates the point.
51. n. 24.
52. Ibid.
53. At present astronauts of seven nationalities are stationed at the ISS. They are involved in operations and experiments undertaken individually and in collaboration.
54. OST assures "a basis of equality" to all states. Refer Article I.
55. *Vienna Convention on Diplomatic Relations, 1961.* Article 4.
56. Refer Article VII (b).

3

Space Policy of India: A Few Pointers

Introduction

India, in space activities, has travelled historic milestones. It reached the moon with Chandrayaan-I in 2008 and is ready to repeat its performance with Chandrayaan-II in 2014. It is planning to launch manned flights in a couple of years and is nurturing the ambition to conduct scientific probes of Mars pretty soon. Other historic achievements are socio-economic development through satellite applications that present a role model to the world and the planned operation of the GAGAN (GPS Aided Geo-Augmented Navigation) navigation system to be executed with a network of 24 high power transponders. And, of course, India's remote sensing satellites are of a high quality, comparable with the best in the world. India, thus, is not simply a space-faring nation but a space power with technological standing and strategic clout, ranking close to the US, Russia, China, European Space Agency (ESA) and Japan.

All these accomplishments have blossomed from India's seminal decision to plunge into space activities that was taken in the early Sixties. It was then indeed a plunge for India had no advanced technology, no developed infrastructure nor trained manpower. The decision emanated from the shared vision of Pandit Jawaharlal Nehru, the first Prime Minister of India, and Vikram Sarabhai, a reputed space scientist of his time. Their wisdom and foresight led to the setting up of the Indian National Committee for Space Research (INCOSPAR) under the aegis of the Department of Atomic Energy (DAE) in 1962 with a mandate to coordinate a space policy, initiate relevant space activities and superintend space research in the country. All these aspects then were in the rudimentary stage and had to be cultivated and fostered. The task though elementary, appeared tall and daunting.

The space programme in India was institutionalised with the formation of the Indian Space Research Organisation (ISRO), again under the Department of Atomic Energy in 1969. And, in 1972, the Government of India constituted the Space Commission for policy formulation as also to oversee implementation of the projects. Considering the importance of space activities and to keep a direct executive vigil, the government established the Department of Space (DOS) to execute parliamentary directions and policy decisions through ISRO and other research laboratories and technology centres. At this stage, ISRO had a homecoming and was rightly brought under the umbrella of DOS from the DAE. ISRO has since grown into a big multi-dimensional organisation controlling sophisticated infrastructure, subaltern organisations and a wide network of support facilities. It also has a commercial arm called Antrix (a Sanskrit word meaning outer space) for techno-commercial consultancy and global vending of space services.

It can be averred at the very outset that India has not promulgated any space doctrine nor declared an explicit space policy nor has the government presented a White Paper on the subject. Perhaps, this was considered too pretentious for a struggler in space technology, that had barely breached the limits of the sky with sounding rockets. The humility was genuine. Hence, no long-term goals were specifically envisioned in the early years nor any roadmap defined for the journey to outer space. It was ad-hoc and largely success-driven but avowedly for civilian purposes, intending economic uplift and social benefit to its large populace below the poverty line.

The above view gets a qualitative endorsement from a recent scholarly study of experts on space security that states, "India's focus has been entirely on civilian applications for social and economic development with very little attention being paid to leveraging space assets or technologies for security or strategic planning."[1] It, in fact, recommends "...a proactive approach to evolving a comprehensive space policy for India ....to best serve India's national security interests,"[2] given the possibilities of dual use of space technology. Earlier, K. Kasturirangan, the then Chairman of ISRO, had also hinted that though India's space assets were solely devoted to civilian use consistent with current state policy yet high resolution imagery has obvious military utility to attenuate national security concerns.[3]

The Evolution of Space Policy

India's Early Interest in the Cosmos
India had exhibited interest in outer space since prehistoric times and its scholars researched and discovered many realities about the universe long before the modern scientific discoveries and abstract theorisations could prove or establish them. Research in space sciences in India is very ancient going by the recorded works in astronomy. The description of various planets and phenomena in the sky is found in the Vedic texts. The contribution of an ancient Indian sage and astronomer, Aryabhatta (born in 476 AD) is significant in that he is credited with the seminal view that the stars did not revolve around the earth. He observed that the earth was round and rotated on its own axis and, thus, caused day and night. He revealed that the moon was dark and shone because of the reflection of the sun's light. He also clarified that eclipses were caused by shadows cast by the earth and the moon.[4] These revelations were remarkable and brilliant for his time.

Again, Sikh scriptures contained in the Sri Guru Granth Sahib[5] are explicit on the subject. The Holy Book asserts about the nature of the cosmos, the estimated population of galaxies, stars and other planets; its origin with the big bang or *Brahm Naad*, its expansion at critical speed and its continuation, the movement of celestial bodies and even the manner in which it shall come to an end to repeat another cycle of existence with a new design and a novel format. These intuitional iterations from divine comprehension by the Gurus and sages have stood firm on comparable scientific validity and the results of modern-day empirical observations.

In recent times, astronomy and allied research into outer space using telescopes and other instruments came to India with the establishment of the Madras Observatory in 1792 and this facility got a boost when the Kodaikanal Observatory was set up in 1898. In modern terms, Indian activities in space sciences can perhaps be traced to the establishment of an observatory in Colaba and the India Meteorological Department (IMD) at Bombay in 1823. About a century later, soon after the ionosphere was discovered, studies of the upper ionised regions of the earth's atmosphere were started at Calcutta University. Soon, research in the field of physics of the upper atmosphere and astro-physics spread to other centres of learning in India. By the late Fifties of the last century, India had developed a strong base for studies of near earth environment and other disciplines related to space sciences with ground instruments and in balloons rising up to 40 km. But true impetus for space activities came in 1969 with formation of ISRO and thereafter it grew in both pace and sophistication.

The Earliest Articulation of Space Policy

The speech by Dr. Vikram Sarabhai on the occasion of the inauguration of the Thumba Equatorial Rocket Launching Station (TERLS)[6] in 1968 is one of the earliest articulations with the semblance of a space policy of India. It was a historic event and he said, "There are some who question the relevance of space activities in a developing nation. To us, there is no ambiguity of purpose. We do not have the fantasy of competing with economically advanced nations.... But we are convinced that if we are to play a meaningful role nationally and in the community of nations, we must be second to none in the application of advanced technologies to the real problems of man and society, which we find in our country."[7]

It seems pertinent to allude to another assertion by Dr. Sarabhai in his address as Scientific Chairman of the UN Conference on the Exploration and Peaceful Uses of Outer Space (COPUOS) at Vienna on August 14, 1968. He said, "I believe that several uses of outer space can be of immense benefit to developing nations wishing to advance economically and socially.... It is necessary for them to develop competence in advanced technologies and to deploy them for the solution of their own particular problems, not for prestige, but based on sound technical and economic evaluation involving commitment of real resources....Indeed, they would discover that there is a totality about the process of development..."[8] The message reiterates the strong linkage between community development and peaceful uses of space technology.

Threshold and Progress

Technological and Intellectual Resources

India started work in space technology almost from scratch and had to almost reinvent its own wheel. It never enjoyed the luxury of transfer of state-of-the-art space technology from the US and the developed West. A victim of many commercial embargoes and technology denial regimes, in a discriminatory denouement, India has seen helplessness from close distance and then made a determined leap of faith on self-reliance. Therefore, the mainstay of India's "technolution" in the space arena has been genuine innovation, creative improvisation, indigenous research and reverse engineering.

It is heartening to note that India has since developed the requisite launching rockets, satellite technology, system integration and fabrication capabilities, and today, many Indian built satellites have been successfully launched with indigenous launch vehicles. These communication and remote-sensing birds[9] have been in operation and continue to do so efficiently and efficaciously in outer space. These satellites are contemporary in technology and rank high in performance on global benchmarks. Their reliable and sustained functioning in the space environment stand testimony to the excellence achieved in the country. Of course, sometimes its trial and error efforts have failed yet initial successes have augured well. Of late, success has become a habit with ISRO. India has, therefore, long since crossed the stage of take-off in space technology and has started soaring high.

Moreover, despite the brain drain to the Western world, India has mustered an adequate pool of qualified and research-oriented scholarship for technology development in its many space organisations. This manpower was motivated and dedicated to breach the techno-barriers and make many a significant breakthrough. Even today, India has a vast resource fund of suitable manpower for the tasks of innovation and improvisation apart from its industrial requirements. Fortunately, India may never face any dearth on this front. But it has missed an opportunity to participate in the International Space Station (ISS) which could have yielded a cutting-edge advantage of trained astronauts ready to conduct scientific experiments on the planned manned flights.

Economic Constraints

It is a universal phenomenon that financial resources are seldom enough to satisfy all, and fully. But given the colossal task of nation building after a long period of colonial rule, India's public kitty has never been able to boast of sufficiency of funds and this has been the perpetual bane. And it applies to the Department of Space in equal measure where, almost always, crumbs were made available to make the best of it. Therefore, most of the efforts in space areas have been rather slow and jerky, thus, causing plan overruns, temporary suspensions and stretched targets. This predicament has, at times, led to a strange paradox of obsolescence of technology, even while development was in incubation, to the detriment of sustained and steady research and continuous progress in mission accomplishments. Sometimes, promising projects had to be aborted or temporarily suspended for the same reason.[10]

The validity of this constraint can be buttressed by the fact that the budgetary allocations for the Department of Space for the first 15 years, since its inception, totalled less than $1 billion while the budgeted amount for the year 2001-02 could barely touch $400 million.[11] Then came the realisation of the accrual of societal and economic benefits and it elicited a spurt in budgetary allocation that was pegged at Rs. 3,499 crore ($ 0.8 billion) for the year 2008-09 and has been raised to an all-time high of Rs. 6715 crore ($ 1.37 billion) in the Budget Estimates for the current financial year (2012-13).[12] India's fund availability, when compared with the budgets of space powers or some other space-faring countries, appears a petty amount.

Let's not unduly bemoan this constraint because India shall almost always, at least in the foreseeable future, face a crunch on financial resources. In fact, there is a lobby in India that questions the fundamentals of space activities in a developing country. This legion of critics maintains that hunger, malnutrition, illiteracy and unemployment remain important challenges and the government would be well advised to prioritise state expenditure for amelioration of the masses. Space programmes in India have a weak constituency and their benefits are neither correctly understood nor politically espoused and still less widely acknowledged. Nevertheless, a judicious balance and a nuanced approach has to be maintained between competing demands though paucity of funds for space research and concomitant applications is likely to remain a constant feature for quite some time.

Formal Iteration by the Government

The crux of the latest space policy can be culled out from the Annual Report of the Government of India.[13] Though purported to be policy, it seems a sheer listing of space activities, broadly generalised, — past, present and future continuous. It is random and not even a prioritised enumeration of tasks. It is brief and cryptic and hardly reflects political vision or scientific foresight or strategic concerns. May be this is intentional. It is, however, euphemistically called the Citizen's Charter of the Department of Space.[14] It directs that "Department of Space (DOS) has the primary responsibility of promoting the development of space science, technology and applications towards achieving self-reliance and facilitating in all round development of the nation". With this basic objective, DOS has evolved the following programmes:

- Indian National Satellite (INSAT) programme for telecommunications, television broadcasting, meteorology, developmental education, societal application such as tele-medicine, tele-education, tele-advisories and similar such services.
- Indian Remote Sensing (IRS) programme for management of natural resources and various developmental projects across the country using space-based imagery.
- Indigenous capability for design and development of satellite and associated technologies for communications, navigation, remote sensing and space sciences.
- Design and development of launch vehicles for access to space and orbiting INSAT, IRS satellites and space science missions.
- Research and development in space sciences and technologies as well as application programmes for national development.

From the above adumbration of "primary responsibility," the specific objectives have been derived and elucidated. Yet it only avers that:

> Department of Space is committed to [the following objectives][15]:-
> - Carry out research and development in satellite and launch vehicle technology with a goal to achieve total self-reliance.
> - Provide national space infrastructure for telecommunications and broadcasting needs of the country.
> - Provide satellite services required for weather forecasting, monitoring, etc.
> - Provide satellite imagery required for the natural resources survey, management of natural disasters, public good services and monitoring of the environment in the country.
> - Provide satellite imagery and specific products and services required for the application of space science and technology for developmental purposes through the central government, state governments, quasi-governmental organisations, Non-Governmental Organisations (NGOs) and the private sectors.
> - Undertake proof of concept demonstration of space applications.
> - Promote research in space sciences and development of applications programme as per national needs.

Apart from the abovementioned policy directives and specific objectives, the following auxiliary goals and subaltern activities appear incidentally entrusted to be undertaken in the national interest, for developmental purposes and for economic reasons. These are spelt out in the Annual Report, 2011-12 as:

> - While implementing the above objectives, Department of Space will:
> - Provide the required satellite transponders and facilities to meet the communications, television broadcasting and security requirements of our country.
> - Provide adequate earth observation capability in spectral, spatial and temporal domains.

- Provide launch services to meet national requirements and commercial needs.
- Provide its products and services in a prompt and efficient manner to all the users/clients.[16]

The Futuristic Vision

Dr. K. Radhakrishnan took over as Chairman of ISRO in 2009.[17] In his media interview[18] on assuming incumbency, he gave some indication of the future trend and his vision of space activities that can be rightly deemed as the current space policy, because it has been vocalised by the top leadership. Some salient points made are in the succeeding paragraphs.

"First and foremost, the space program today is integral for the country's development. India, today, is a role model for the world on peaceful applications of space research."[19] This implies that India remains committed to its primary aim of socio-economic development through beneficial civilian uses of space technology. The existing thrust on this front deserves to be maintained and, if possible, further accelerated and diversified.

Referring to the ongoing projects, he mentioned the launch of the Geo-Synchronous Launch Vehicle (GSLV) Mark3, which will take India to the four-ton launch capacity. For this project, India will be using an indigenous cryogenic engine for the first time. The launch will also enable testing of several critical technologies like solid-fuel strap-on motors with 200-tonne propellant, and liquid stages with 110-tonne propellant. The engine is being integrated on the launch pad. The success of this project is crucial to the efforts and morale of the space science fraternity as well as its lucrativeness for commercial projections to customers with heavy payload satellites.[20]

Success, indeed, has a heady aroma and consistent achievements have bolstered the confidence of Indian space scientists. They have started dreaming beyond the confines of a tame and modest space policy. In the same vein, Dr. Radhakrishnan has referred to the project GAGAN (GPS-aided Geo-augmented Navigation) System, which is India's own version of the Global Positioning System (GPS) restricted to the territory of India with marginal coverage at the contiguous borders. The system will work on a network of 24 transporders in the Ku band and is expected to interface with the GPS, GLANOSS and Galileo systems. It is a satellite-based navigation system for which a dedicated payload will be launched on GSAT-4. The Final Operation Phase (FOP) of this system was initiated in June 2009 and was expected to be ready for testing during 2010-11.[21] It has since become operational. This project of the National Positioning System, in all probability, may have military implications also.[22]

Alluding to the accomplishments in the area of remote sensing, Dr. Radhakrishnan claimed that India was "on par with other global powers in the field. But we still need to develop satellites that look at the atmosphere, study cloud movements, etc. Our Cartosat-2 ranks among the best in the world with a 0.8 metre resolution camera on board. It is capable of providing scene-specific imagery and various cartographic applications at cadastral level. The system is functioning well."[23] The next satellite in the carto-series will be the Cartosat-3, scheduled for launch in 2012-13.[24]

A prestigious project is the follow-on experiment with Chandrayaan-2, which will launch a rover on the moon for exploration in space sciences and mapping of mineral resources. The space-vehicle will have three parts, Orbiter craft, Lunar-Lander craft and Rover craft, to be developed under an Indo-Russian joint venture. In this project, India develops the Orbiter that will launch and carry the module up to the lunar transfer trajectory and thereafter orbit around the moon at about 100 km. India is also responsible for developing the Rover craft, equipped with robotic arms, for movement on the surface of the moon and with facility for excavation of rock samples. Russia is to undertake development of the Lander craft for a soft landing on the moon. This project was slated for launch in 2011-12.[25] ISRO has since finalised the payload and scientific instruments for the mission but has postponed the launch to early 2014.[26]

Another ambitious programme is the first manned space flight mission. The ISRO proposal relates to sending two Indian astronauts on board a space capsule on a low-earth orbit of around 400 km that will keep encircling the earth for about a week. Regrettably, India missed the chance to train a core team of astronauts through participation in the ISS experiment. India failed in its vision of future space activities.[27] It was a policy blunder. Though the ambient environment and space allocation inside the craft as well as other parameters are standardised, the India-specific requirements may cause compromises. This critical mission would, of course, demand the same level of preparedness as the moon mission. It is planned for 2015-16.[28]

Space tourism is the industry of the 21st century and harbingers of its phenomenal growth are already evident. Russia has taken the lead in commercial space transportation and other players, like Virgin Galactic and Space Exploration Technologies, are ready to join the race. India, too, is keen to capture a share in this market. Towards this end, "a series of technology demonstration missions have (sic) been conceived [and] for this purpose, the Winged Reusable Launch Vehicle (WRLV) technology demonstrator has been configured....and [its] aerodynamic characterization....completed"[29] The demonstrator will act as the testbed to evaluate various techniques viz. hypersonic flight with or without air breathing propulsion, autonomous landing and powered cruise flight. First in the series of demonstration trials is the Hypersonic Flight Experiment (HEX).[30]

Another major development that remains consistent relates to the mini-satellite concept.[31] In this project, the Indian Mini Satellite (IMS-1) weighed 83 kg and had miniaturised sub-systems. It was flown on board the Polar Synchronous Launch Vehicle (PSLV-C9) for remote sensing applications during 2008-09. The data from this mission is being made available to the interested space agencies and the student community from the developing countries for capacity-building in using satellite data.[32] The second satellite in this programme is named YOUTHSAT and was launched as an auxiliary satellite in the polar synchronous orbit on April 20, 2012,[33] with a designed life of two years. It is part of the Indo-Russian collaboration for scientific studies of the terrestrial upper atmosphere and measuring of solar radiation. For this purpose, this satellite will hoist payloads of scientific interest with the participation of the youth from universities and research scholars to provide them an opportunity and to inculcate interest in space related activities. Incidently, ISRO successfully launched a PSLV on May 9, 2010, that carried the StudSat (acronym for student satellite) of the pico satellite class, weighing 850 gm, and designed by undergraduate university students.[34] The ultimate plan is to operate a micro-satellite bus carrying different kinds of payloads for varied applications and scientific experiments.[35]

Indian scientists and engineers are bracing up to launch an average of 10 satellites per year to meet the rising demand for various space applications, including communication and remote sensing.[36] The calendar of activities for the current year is hectic and fully packed with surprises. ISRO Chief, Dr. Radhakrishnan, has been quoted as saying that India could be part of the global human flight to Mars.[37] Thus, manned missions to the moon and winged reusable space vehicles are passé.[38] India is getting into the league of the space powers, with plans truly futuristic and highly ambitious.

Creditable Achievements in Space Technology
First, India may have started from a low ebb in space technology but it has achieved some creditable firsts in space activities and discoveries. The Chandrayaan mission has been a singular success of remarkable significance. This probe carries the honour of sending data that undisputedly established the presence of water on the moon. It has, thus, hurtled India into the rank of space powers and credited India with pioneering global achievements in space technology. Secondly, with the successful blast-off of the PSLV-C11, India recorded the thirteenth consecutively successful launch to prove consistent reliability, operational versatility and soundness of design. Thirdly, the PSLV-C9 that launched the CARTOSAT-2A on April 28, 2008, created a world record of launching ten satellites as co-passengers in a single flight. Apart from the advanced remote sensing satellite, it carried the Indian Mini Satellite (IMS-1) and eight nano-satellites for international customers. As on November 2011, the Indian constellation in outer space comprises eight communication satellites, two meteorological satellites, ten earth observation satellites and one science satellite.[39]

Fourthly, the CARTOSAT-2B, successfully launched with the PSLV-C15 on July 12, 2010, competes globally and offers services second to none in the world. And further, StudSat, launched on the same vehicle, was a fully integrated and operational satellite that was designed by university students. It belonged to the pico-satellite class, weighing only 850 grams. The same launcher was also used to hoist one Algerian, one Canadian and one Swiss satellite.[40] An updated report on the mission confirmed placement of satellites in the respective orbits, beaming of hi-resolution images and receiving of clear signals from StudSat.⁴¹ The success is heady. India fervently hopes to continue in the same stride and strive for still better results. Fifthly, redeeming the pledge of self-reliance in space technology, activities for the realisation of the GSLV Mk-II and Mk-III for the two-ton and four-ton class respectively, with the indigenous cryogenic engine, are progressing satisfactorily and on schedule.⁴²

Sixthly, the GAGAN system has since become operational. For this purpose, an advanced communication satellite, the GSAT-8, carrying 24 hi-power transponders in the Ku band and the GAGAN satellite with two channels in the L-band were placed in a geo-stationery orbital slot of 55 degrees East longitude. The transponders have augmented the capacity in the INSAT system while the GAGAN payload provides the satellite-based-augmentation system through which the accuracy of the positioning information obtained from GPS satellites is improved. Its range covers the Indian mainland and the Andaman and Nicobar Islands. All deployments were successfully carried out and the satellite was 3-axis stabilised. Orbital testing was done in June 2011, and fine-tuning is in progress; and its system certification is expected by July 2013. Thereafter, the Navigation Satellite System will be declared operational.⁴³ Despite speedy efforts, the system is expected to become operational for aerial navigation of commercial aircraft by late 2013 or early 2014.⁴⁴ It has power generation capacity with a mission life of more than 12 years. With the GAGAN system becoming operational, India would join the exclusive club of the US, Russia and the ESA. China may also barge in soon with its Beidou system.

Lastly, India operates the biggest and widest socio-economic developmental network in the world, with a variety of space applications. This focus has not shifted and continues to be an inalienable strategy for societal welfare and for the uplift of the masses. This view is acknowledged by NASA that has identified "India as a 'big partner' of the US space programme". NASA has applauded the country's efforts in using space missions for "societal needs"⁴⁵. The applications encompass telecommunication, broadcasting, tele-medicine, tele-education and rural development. The goals are laudable and have been commendably achieved to a great extent.

Lately, with the Gramsat, the scope and footprint of activities has been broadened so that info-centric benefits can permeate to the grassroots level in the districts and ultimately to the villagers through Village Resource Centres (VRC) being set up all over India.[46] Similarly, the tele-education network has more than 55,000 classrooms and the tele-medicine programme facilitates connectivity of 382 hospitals to 306 rural hospital and 16 mobile vans to 60 super-speciality hospitals providing health care to the citizens.[47] The hallmark of the Indian space programme has been to ensure societal services by INSAT satellites that are oriented toward benefits to a large segment of the deprived rural population in commitment to the true ethos of mass development.

New Strategy Mandates

Strategy is a derivative of policy but sometimes the tail wags the dog, meaning that strategy influences or dictates policy and this is happening here. India has come of age in the space sector and has been acknowledged not only as a space-faring nation but also as space power by its own prodigious standing in technological research, vehicle launch record, footprint of applications, integration of satellite systems, deep outer space probes, control and monitoring of satellite operations, remote sensing capabilities and allied expertise in space activities. India has literally emerged as a space power and is accordingly reviewing its achievements in human development applications, assessing its peaceful initiatives in geo-spatial data acquisition, realigning its objectives based on the threshold of technology and redefining its strategies in consonance with geo-political realities and concomitant national security concerns. Consequently, some emerging mandates are discussed in the succeeding paragraphs.

Space Applications for Socio-Economic Development

India embarked on its space programme with the avowed objective of providing benefits to its teeming masses and has been stupendously successful in this task. In fact, India has emerged as a role model for other developing countries in this respect. This task so commendably accomplished should not be relegated or abandoned: on the contrary, it should be encourage to diversify in approach and content and, at the same time, should strive to extend its reach and widen its footprint for the greater benefit to the larger population. This strategy should be sustained for progress.

India can today boast of an impressive array of achievements and mastery in diverse areas of space technology. The first and foremost relates to telecommunication and television broadcasting with allied applications for tele-education and tele-medicine. The composite effect of all this has been sustained national socio-economic development and alleviation of poverty and illiteracy, and public health services. The excellent performance of INSAT-2A, since its commissioning in August 1992, provides indisputable testimony to indigenous capability for building world-class multi-purpose satellites under the unique Integrated Mission for Sustainable Development (IMSD) scheme.

Over the subsequent years, satellites in this class like the 2B up to 2E and INSAT-3 and 4 series have progressively carried a more sophisticated and state-of-the-art payload for broader utilisation in purpose and footprint. They have also been providing advanced meteorological data and vital services in disaster warning relay and distress alerts. Therefore, emphasis on societal developmental applications, initially envisioned for civilian socio-economic causes, is crucial; and should be consistently continued, reinforced and diversified as much as possible.

Scientific Probes in Service of Humanity

The Chandrayaan-1 experiment, that cost $82 million, has succeeded despite the premature demise of the Lander on the moon. It has, in its lifetime, yielded good results and useful data. A major find has been of the presence of water on the moon and this scientific confirmation has been globally acclaimed. As a sequel to this, Chandrayaan-2 is in the offing as a joint venture with Russia. Apart from other experiments, the robotic arm of the Lunar Rover shall excavate and collect soil samples for on-the-spot testing and transmit data via the Orbiter to the earth station for analysis. The launch was scheduled for 2011-12 that has now been postponed to early 2013. It appears an ambitious project but realistic in achievement.

With expertise gained in space ventures, India is poised to harness technology for more complex inter-planetary missions, deep-space exploration, proof-of-concept experiments and furtherance of space sciences. The accomplishments should not tempt for grandeur or prestige, nor even for hegemonic ambitions, but be in the service of humanity with the due humility of an achiever and in deference to the space science sorority. In fact, all this should be in aid, and for the benefit, of humanity, to improve quality of life on planet earth and add to the repertory of human knowledge.

Plans to reach other heavenly neighbours in quest of scientific exploration to satisfy human curiosity and to develop space sciences have been envisioned and projected. "Around [the year] 2015, an unmanned spacecraft should blast off for a rendezvous with Mars. ISRO, buoyed by a string of successes, is evaluating launch opportunities for a low-cost voyage to the 'red planet'. Only the foolhardy would be willing to bet on its failure."[48] The scientific vision is clear and the focus unwavering yet the imponderable is whether the government shall make adequate funds available for the purpose that could cost $81.6 million. The governmental clearance is still to be received, yet according to a knee-jerk political statement on the occasion of the Curiosity Rover landing on Mars on August 6, 2012, this mission will now be preponed to October/November 2013. This seems unrealistic yet the intention could be to keep abreast with Asian competitors, namely, China and Japan. The dilemma still exists, yet the future beckons.

Emerging Security Imperatives

It would be incorrect to assert that India has not yet mulled over the thought of space-based advantages for national security and defence imperatives because, in contemporary times, space capabilities are getting embedded into the security and war doctrines of space-faring nations globally. In fact, military spin-offs of civilian space programmes are increasingly becoming obvious because of the dual uses of technology. Therefore, it would not be unrealistic to surmise that Indian space-strategists and scientists have already proposed, theoretically demonstrated and sought funds for, proof of concept experiments. Perchance, if such considerations have not visited their heads to excite a futuristic approach, it would seem highly pertinent to catch up on time lost or opportunity frittered away. A serious thought process needs to be devoted to this aspect because space is no longer an esoteric medium and the military relevance of this domain is becoming increasingly germane to victory in war.

India's progress in space activities has been spectacular and the developmental uses phenomenal. The military spin-offs of these achievements through dual-use technology are obvious and feasible. There are three basic characteristics of military space systems. First, in terms of vantage, space is the ultimate high ground and earth-observation satellites offer an unprecedented vantage in surveillance and reconnaissance of borders and battlefield. Secondly, as for pervasiveness, space assets are free to operate without national boundaries or weather perturbations and this ability allows unprecedented access to any area on the battlefield. Lastly, regarding timeliness, space assets can provide continuous or periodic access over areas of interest and more timely information than is generally available from terrestrial systems. The combined impact of these advantageous factors can be synergistic and function as a significant force multiplier.[49]

To eke out the above advantages, the adaptation of the existing national space apparatus for civilian uses can be the starting point and threshold for specificities. The defence applications achieved by other space powers can provide the footprints to follow, and the lessons learnt from experience by the space pioneers can provide the roadmap to progress. However, the first step in achieving a viable military space capability is to articulate a new direction for space policy and introduction of this concept in military doctrines with detailed modalities. This has to emanate from a clear sense of purpose and with determination at the highest level. Needless to reiterate that capabilities exist, the requirement is real and the potential market responsive; as a result, the threshold is ready for a tryst. Moreover, this is a long gestation activity and its fruition would take quite some time, hence, timely decisions are crucial.

The requisite possibilities, even at current level of assets, diversified activities and the threshold of experimentation, may seem tempting to yield tantalising opportunities and promising results too easily, too soon. There is growing clamour to acquire and operationalise military applications, at least passive ones.⁵⁰ More so, in view of the proven utility of these facilities for reconnaissance and surveillance, real-time theatre-operation information with attendant force multiplier effect, communication connectivity and in guidance of bombs to minimise unintended collateral damage on the targets during the recent wars in Iraq and Afghanistan.

The advantages seem many and myriad, with temptations too strong to resist. Military commanders and defence strategists have been prodding the government to abdicate self-reclusion and shed doctrinal orthodoxy.⁵¹ The government may well recognise the professional urges and legitimate projections of the operations commanders. However, India has resisted internal pressures and abided by its national pledge of space for development and has persistently opposed weaponisation of outer space by repeatedly and vehemently so asserting at the Conference on Disarmament (CD), at the Prevention of Arms Race in Outer Space (PAROS) discussions, and in other world forums.

Another area of security pertains to protection of national assets in space and eventually on the celestial bodies. This assumes importance as India has a large constellation of space satellites⁵² in operation on vital missions providing crucial facilities and scientific data. Further, excavation activities on the moon would need local safety measures. India can ill-afford any harm, disruption or attrition. Therefore, to safeguard assets, space-mines comprise just one option⁵³ among others. "Therefore, India needs to have a well-calibrated Outer Space Policy that...factors in the capabilities in outer space that are vital for protecting security concerns."⁵⁴ In a nutshell, India needs to reinvent itself and redefine its space fundamentals as well as make optimal utilisation of its space assets to address genuine security concerns.

Carving a Commercial Niche
Space, today, is a booming industry and business prospects in activities relating to uses of space are growing exponentially. Driven by the telecom, TV and internet revolutions, the global space market for satellites, launch vehicles, and earth-based facilities was expected to rocket to nearly $320 billion by 2010 AD.⁵⁵ The software for the growing space infrastructure will involve even larger sums. As of today, about 500 birds are in orbit supporting

telecommunications, remote sensing, TV, scientific experimentation and military-applications. In the next decade, a galaxy of satellites will roam in outer space and business prospects will open up new vistas for burgeoning opportunities and diversified callings.[56] The US has already outsourced the shuttle taxi service to private enterprise and other less sensitive tasks may follow. India can strive for a slice of this pie in its known expertise.

India has gained expertise and experience in satellite fabrication, systems integration, software development, launch vehicles, remote sensing and data acquisition facilities. Particularly, in the areas of remote sensing and geo-spatial monitoring, Indian satellite facilities rank equal to the global best and are well sought after. India has carved its own niche of specialisation and it could materialise great business if the sales thrust is professional and competitive in the space business market. Possibly, India can leapfrog its competitors in cost-economics and timed-deliveries. Therefore, a commercial space policy for the 21st century must not be about passively protecting the existing limited advantage but striving aggressively to reach for the substantial gains of emerging flanking technologies, new fangled industries and commercial endeavours in the space-faring future scenario.

Another opportunity arises because space programmes involve high costs, have imponderable risks, long gestation periods to fruition and, due to the global dispersal of expertise in space technologies, a marked trend is discernible where future projects in outer space are expected to be joint ventures, bilateral and multilateral. Thus, international cooperation becomes the central theme of space science development. This trend would necessitate a contingent shift in the space policy of many space-faring countries whether to outsource space modules or resort to shelf-shopping of products and facilities on the space mart. Both options appear advantageous.

Today, the US, Israel, Germany and other space powers are engaging in commercial cooperation with India and others to partake of the accrued reciprocal benefits. For example, Indo-US relations are visibly improving due to the civilian nuclear deal of 2008 and greater economic interaction is bound to be mutually advantageous. Perhaps, in view of India's track record in space probes, the US may profitably outsource non-crucial, non-strategic and non-sensitive products and activities to save domestic resources for more sophisticated and sensitive projects. Such transactions would sure lead to auto-catalytic development of space technologies and synergise expertise.

India has, in fact, started offering space hardware, launch facilities, software applications and consultancy services through Antrix, the commercial arm of ISRO, established in 1992, on the global space mart. Creating off-the-shelf availability of space products like mini/micro/ pico class of satellites, offering low-cost multiple satellite launch facilities and technical consultancy to new entrants into the space arena would make good economic sense. During the year 2011, this corporation has done good business with commercial contracts for the launch of X-Sat of Singapore and VessalSat-I of Luxembourg on board the PSLV-C16 and PSLV-C18 respectively. The contracts have been fully executed.[57] Antrix has signed another commercial agreement with M/s. Astrium-SAS for the launch of a remote sensing satellite, SPOT-6, during 2012.[58]

Fortunately, Indian products and services on offer are relatively cost-effective, reasonably high-end on state-of-the-art technology and can match the global benchmark on quality. Added to this is the Indian experience and expertise in space launches that is amply reflected in its success rate. Indian space products and facilities are available through Antrix that operates as a corporate entity and has taken on this specialised task of marketing in the space segment. It is doing well but not well enough for kudos. It needs to be market-savvy and grow professionally, the sooner the better.

Cooperation in Support of Self-Reliance

Fully conscious of its technological limitations, India has always tried to seek cooperation in space activities from wherever possible and has been keen to hold any willing hand in friendship for gaining technical solutions. As the wise say, one does not look a gift horse in the mouth. Following this adage, India welcomed and assimilated whatever bits of space technology came its way to fill gaps in its indigenous efforts. This policy was pragmatic though it belied complete independence and, by inference, betrayed a weak international profile. Nevertheless, the spirit of cooperation succeeded well to augment its fund of technology and provide fillip to its knowhow. Further, despite formidable in-house research and massive development effort within the family of institutions paternally headed by ISRO, India still faces a technology deficit. All round efforts are fully devoted to overcome this.

On the domestic front also, India's space programme never lost focus on self-reliance and sourcing of indigenous capabilities. For this purpose, an extensive plan was chalked out for cooperation with academic institutions, public sector undertakings and private industry. The initiatives have worked well to supplement in-house research in satellite design as well as launch facilities and fabrication capabilities of ISRO and allied organisations under its lead. The domestic effort so harnessed has been stupendous and these robust initiatives towards much needed technological breakthroughs ought to be evaluated, extended and expanded.

ISRO elicited academic involvement in technology research through an initiative called RESPOND to encourage academia from universities to participate and contribute in the development of relevant areas of space science and technology. This programme started in the early Seventies and selected national institutions and universities are doled grants-in aid for specific research in different functional areas. For ensuring more active involvement, ISRO later set up Space Technology Cells at premier educational institutions of higher learning to coordinate activities in space technology and applications. It has, thus, established strong linkages with academic institutions to derive useful outputs of research and development. It has also enhanced the base for generation of human resources and strengthening of infrastructure to widen the catchment area.

Similarly, for the involvement of the corporate sector and private industry into space technology and applications, ISRO evolved a multi-faceted programme and pursued it as a conscious policy for nurturing the essential infrastructure of the vendor market in telecommunications, broadcasting, meteorology and natural resources management. To synergise industrial capabilities to maximally support the infrastructure, ISRO infused transfer of technology for absorption, utilisation and innovation in product manufacturing. Fortunately, the industry seized the potential commercial opportunities and responded with enthusiasm to meet the impending challenges. Promising results have started flowing in.

India's resolve to be self-reliant in space technology was borne out of circumstantial compulsion of gross denials, arbitrary sanctions and a discriminatory licensing regime on transfer of technology by the Western world due to the perceived nuclear ambitions of a developing nation. The embargoes operated across the board and even on simplistic dual-use technologies. Thus, India's technological isolation was near complete and even peaceful initiatives were adversely affected or hampered. Hence, it was out of sheer necessity and pragmatism that India opted for selective international cooperation in space activities and training programmes that has remained a cornerstone of its foreign policy and aided its space projects.

International Cooperation

International goodwill and cooperative relations have been evolved over the years and continue to be important contributors to India's space activities. In illustration can be mentioned that in the Sixties, the Thumba Equatorial Rocket Launching Station (TERLS) was set up with the cooperation of the USA, USSR and France. India opted for the US ATS-6 satellite for the Satellite Instructional Television Experiment (SITE) and the Franco-German Symphonie satellite for the Satellite Telecommunication Experiment Project (STEP) in the Seventies of the last century. The launchers for the Aryabhatta, Bhaskara and IRS satellites were provided by the USSR. APPLE was launched on board the European launch vehicle, Arianne. The Indo-Soviet joint manned space mission of 1984 was used to conduct remote sensing and bio-medical experiments in space while the Viking liquid propulsion engine technology was obtained from France.[59] India is currently receiving microwave remote sensing data from the European Satellite, ERS-1.

It seems pertinent to mention the latest international cooperation pursuits. These are the Megha-Tropiques project in collaboration with France and YOUTHSAT, jointly with Russia. Megha-Tropiques is a meteorological satellite for improving the understanding of climate and weather systems as an exclusive observational system. It was launched on October 12, 2011, and is expected to help generate improved current-weather variables, leading to better forecasts for the entire tropical region. No wonder, "Scientific teams from Australia, Brazil, Japan, Republic of Korea, Sweden, United Kingdom and United States of America are awaiting data from this satellite."[60] Further, the Global Precipitation Measurement (GPM) capability of this satellite is likely to be a boon to the world's "scientific community engaged in research on climate and weather systems that affect the daily life of humankind the world over and particularly in the tropical region."[61]

Again, YOUTHSAT, launched on April 20, 2011, was jointly developed by young scientists from ISRO centres and Moscow State University in Russia. It is "a satellite for studying atmospheric constituents and space weather...solar flare activity, and to detect the bursts of X-rays, energetic electrons, protons and Gamma-rays."[62] Another collaboratory project under progress is Chandrayaan-2 with Russia. This project also seeks cooperation from other countries like Canada. After repeated postponements, it is now planned for launch in 2014.[63]

At present, India has bilateral or multilateral agreements or memoranda of understanding or framework agreements for formal cooperative arrangements with many countries. The prominent in the long list are the US, the UK, Canada, China, France, Germany, Israel, Russia, Sweden, Ukraine, Australia, Brazil, Japan, etc. Apart from these countries, cooperation instruments exist with space agencies like the ESA and space related bodies like the European Organisation for Exploration of Meteorological Satellites (EUMETSAT). Thus, through international cooperation, India "continues to take up new scientific and technological challenges; refine space policies and define an international framework for exploitation and utilisation of outer space for peaceful purposes."[64]

Another aspect of India's cooperation extends to helping the Third World. In recent times, it seems pertinent and important because with rapid technological advancement among space-faring nations and greater awareness among the rest globally, the scope of international cooperation has become wider and diversified. India has always harboured altruistic motives regarding help and aid towards other developing and 'space-fledgling' countries. It has, therefore, sought out areas of cooperation to help those Third World countries that could not, for some reason, indulge in full scale space activities to eke out socio-economic benefits.

For this purpose, India shares its own experience with, and makes available, training facilities to other developing countries by training their personnel under a programme called SHARES (SHARing of Experience in Space). In this direction, India has helped many Afro-Asian and Latin American countries, particularly Malaysia, Mauritius, Syria, Mongolia, Hungary, Algeria, Singapore, Thailand, Myanmar, Republic of Korea, Argentina, Chile, Venezuela, Peru and several others. India has been enriched by this bilateral cooperation while honouring the realities of the big brother relationship in the region.

On the regional scene also, India is currently working with the Secretariat of the Association of Southeast Asian Nations (ASEAN) to make available data from IRS satellites to all the ten nations for disaster management. It is also negotiating with the South Asian Association for Regional Cooperation (SAARC) to establish a network of weather stations to support thunderstorm predictions and severe weather warnings. Similarly, in the Asia-Pacific region, too, ISRO continues to contribute to the activities of the Asia-Pacific Regional Space Agency Forum (APRSAF) through sharing expertise for the benefit of this region.

International Implications

Lately, in the quest of a global profile, India has been attempting to recast its relations in the international arena based on congruencies in geo-political imperatives and shared concerns in outer space to establish synergistic catalysts. This has significant implications for its strength and standing in the global listing. Another parameter that has shaped the quality of affiliations has been technical help and cooperation received or denied in space technology. No doubt, self-interest has prominently dominated to crystallise its global interactions.

India has also been actively participating in the UN Committee on Peaceful Uses of Outer Space. Prof U.R. Rao, an eminent Indian scientist, headed the IIIrd Conference on Exploration and Peaceful Uses of Outer Space in 1999. India's participation in the UN and other world forums concerning space matters has since become vocal and visible befitting its exalted status as an "almost-space-power". India's relations with some of the space powers and the international implications thereof, with particular reference to the US, are discussed in the succeeding paragraphs.

India and China

India has reached an exciting time in the history of its space programmes that have become a source of prestige. It is a turning point in its space profile that "catapults it into a space race with China."[65] The two countries are generally perceived to be rivals in the space race and competitors in space commerce, even though China has made huge strides in the military uses of space and is galloping much ahead of India and other Asian space-farers. The military orientation of China's space programme is indicated by an eloquent historical fact that Chinese SLVs are a derivative of their ballistic missiles developed with the help of the erstwhile USSR whereas, in contrast, India's SLVs were first produced for the space programme for development applications and have only later been adapted and modified as ballistic missiles.[66] The asymmetry in space technology and incongruence of space missions between the two countries is apparent and obvious.

Of course, India has achieved "threshold capability" and is steadily progressing on the GAGAN national navigation system but China seems well ahead and rather serious about its Beidou navigation system for global positioning. This system ultimately shall have a constellation of five geo-stationary and 30 non-geo-stationary active satellites by 2020.[67] Though, ostensibly, this system "is expected to focus on storm and earthquake forecasting and disaster rescue operations" yet possible military nuances can hardly be overlooked. Thus, China aggressively barges into the US-Russia-Europe cartel of navigation facilities, with the capability to exploit these assets for global eavesdropping and other defence usages such as "information support," "battlefield combating" and "battle-space characterization."

The calculated military slant to China's space programme is no secret nor does it need any evidentiary proof but its promiscuous demonstration of military prowess and hegemonic postures is disconcerting. This tendency can be illustrated by the Anti-Satellite (ASAT) experiment on January 11, 2007. In this incident, China utilised its ground-based missile to shoot down its defunct satellite. Apparently, this experiment was aimed to prove to the US that China possesses effective anti-satellite technology. The Pentagon's citadel of Ballistic Missile Defence (BMD) got breached and it also revealed discernible chinks in the US space security system. It has caused a serious jolt to the US supremacy[68] and, at the same time, has impacted hard to wake India out of its slumber to review the security *realpolitik* of Asia and safeguard its vulnerable flanks. The loud warning should be heeded by all because it has instructive implications relating to the definition of space weapons and their ban in outer space. Another disturbing development relates to China's contemplation of fierce competition for space resources and its irresponsible vocalisation of territorial claims to the moon and outer space that are utterly untenable under the Outer Space Treaty. The Chinese ambitions are ominous.[69] On these aspects, the US has its own reservations, and the Sino-Russian bloc, its own agenda.

In the remote sensing arena also, China has forged ahead of India. "China has launched nine remote sensing satellites in the *Yaogan* series since 2006 using its Long March range of rockets."[70] In comparison, India with the launch of the Cartosat-2B in July 2010, just about matches Chinese capabilities, yet has reached the 'critical constellation' of ten active remote sensing satellites in space. Besides the images beamed by its panchromatic cameras have exhibited hi-quality and hi-resolution. The edge gained by India is too distinct and significant to be frittered away.

Be that as it may, it is well acknowledged that "India's [space] program is smaller in scope than China's and is thought to receive far less funding. It is also designed mostly for civilian purposes, whereas experts have correctly surmised that China is more interested in military applications."[71] In a nutshell, expert opinion veers round to accept that "China is still the leader. India has yet to diminish China's space stature. But India is indeed seeking a higher global profile."[72] The technology gap between them is visibly wide and may widen further, specifically in military applications.

The Chinese progress is stupendous and future plans literally aim for the sky. It would be relevant to allude that China is almost ready with the first module of the Space Station, Tiangong-1 to be joined by two Shenzhou spacecraft over the next two years as part of the manned space programme. The second Lunar probe and an unmanned moon landing are also in the offing.[73] For India, it shall remain a handicap race, with directional mismatch making it equally irrelevant to overtake. Nonetheless, China is wary of the strength and preparedness of India's defence forces and its military deterrence posture. The Pentagon has reported that Beijing is moving nuclear-capable missiles closer to India's border.[74] This unprovoked move may change the threat perception and have strategic repercussions on India's space policy.

The adversarial attitude and rivalry between India and China appear inherent and obvious yet it may surprise many that overtures have emanated from both sides for collaboration in space activities. This augurs a possibility, however slender and remote, of 'Chindia' cooperation that may not seem comfortable either to the US or Russia and may have the potential to tilt the balance of power or shift the centres of economic and military might.[75] The US, of course, cannot brook such skewed polity and perforce has to regain unchallenged supremacy. In one of the possible scenarios, the US may be tempted to court India to counterpoise China's rise in technology and economy. Thus, India may get a windfall.

India and other Asian Competitors

There are many Asian countries jostling to enter the portals of the space league but some of them are mere strugglers and do not reckon for competition. The noticeable Asian competitors in the arena are Japan, South Korea and Pakistan while North Korea and Iran have been intentionally kept out of the discussion. No doubt, all of them are striving for comparable independent and autonomous capabilities but their mission objectives are divergent and, thus, not really competitive.

For instance, Japan has a well developed and multifaceted space programme that is credible and visible. Japan started from scratch but its sterling characteristic of resilience paid off. Today, Japan enjoys dual-use capabilities of its space assets and can boast of potent Command, Control, Communications, Computers, Intelligence, Surveillance, Reconnaissance (C4ISR) capabilities comprising space-based communications, navigation, positioning, targeting and a two-tiered BMD architecture. Japan would soon project itself "as a significant military force to contend with, not only in the Asian continent but in the entire world."[76]

Notwithstanding, Japan's space probes primarily have a scientific orientation that concentrates on planetary explorations and astro-physical sciences. This is borne out by the Japan Aerospace Exploration Agency's (JAXA's) impressive seven-year sojourn by Hayabusa and the planned launch to Mercury in 2014, using a craft that will be covered in mirrors to reflect the heat of the sun while, at the same time, designed to be partly powered by solar energy.[77] It is surmised from diplomatic initiatives that India and Japan may enter into a collaborative relationship with joint ventures for scientific exploration and deep space probes. Both are likely to find each other a trustworthy partner for deeper ties.

South Korea is a late entrant to the club of space-faring nations and has developed its space capabilities with the US' benevolence and generous doles of technology, though with strings attached.[78] It has tried to build its expertise on this baseline but "[t]o make up for lost time, South Korea has adopted a mid-entry strategic approach…to leap-frog to requisite technology levels through technology transfers." This strategy was coupled with a smart selection approach to "select areas of development based upon its unique needs and resources instead of trying to accomplish overall efficiency in all areas."[79] It appears a prudent policy for the country and presents no threat to commercial opportunities for India.

Furthermore, Pakistan offers no real competition to India in this field. In fact, Pakistan's own space programme comprises more turnkey projects proliferated from China, with little indigenous efforts. Reports on Pakistan are replete with authoritative references to covert transfer of technology by China and minor technology chasms bridged with the help of a friendly US. It appears that Pakistan has not assimilated space technology and has always looked to quick-fix solutions from trusted friends. However, borrowed plumes do not make a good spectacle.

"It can be safely surmised that while Pakistan has immense ambitions (both military and civil) with regard to space, it currently lacks the wherewithal for the same" but "Chinese and North Korean collusion may in the near future allow at least a partial consummation of its desired peace capabilities."[30] Another expression of this ambition is reflected in its obsession with leadership of the Islamic world for "promoting the advancement of space sciences and technology in the countries of the Islamic world." With this aim, Pakistan formed the Inter-Islamic Network on Space Sciences and Technology (ISNET) in 1986, bringing together Muslim countries.[81]

Well, empty vessels do make a lot of noise.

India and Russia

Indo-USSR cooperation goes back nearly half a century and was formalised when India signed a Treaty of Friendship with the erstwhile USSR in 1970. They have since forged a durable and dependable mutual relationship in matters of politics, economics and technology. Soviet Russia has always been supportive of India's endeavours into outer space and realised its handicap in space expertise. Therefore, short of direct transfer of technology in blatant disregard of the ethos of the space powers, it has been offering help and facilities to accelerate India's efforts to achieve its targets. Sporadic technical hints and informal bits of information on cryogenic technology have also helped India create an indigenous cryogenic engine by reverse engineering and experimentation. This scientific harvest has indeed paid dividends.

Specific mention of such succour would vindicate the above statement. Initially, when India was nearly isolated from the Western space powers, Soviet Russia offered launch facilities from the Baikanour spacedrome and provided launchers for the Aryabhatta, Bhaskara and IRS satellites. A major landmark in mutual relations came in 1984 when an Indian astronaut, Rakesh Sharma, joined the Soviet crew to orbit around the earth. This joint manned space mission was utilised to conduct geo-spatial remote sensing and bio-medical experiments in space. The results of the data collected were promising, and the crew cordiality was exemplary.

Another prestigious project of cooperation is the planned Chandrayaan-2 experiment, which will launch a Rover on the moon for exploration in space sciences and the mapping of mineral resources of the moon. The space-vehicle will have three parts, comprising the Orbiter, Rover and Lunar-Lander craft which will be developed under an Indo-Russian joint venture. While India develops the Orbiter and Rover, Russia is responsible for providing the Lander craft. The Rover module will excavate and collect soil samples from the surface of the moon for on-the-spot mineralogical analysis. This project was slated for launch in 2011-12 and has now been postponed to early 2014.[82]

India and the Soviet Russia have cavorted for too long and as a result India has started leaning too much on the Russian shoulder. In this regard, the statement of Russian President Dmitry Medvedev made on August 15, 2010, is pertinent. "The strategic partnership with India is an 'unconditional priority' of Russia's foreign policy."[83] This stance seems to assure India of customised support in contingencies. Initially, the US, having ushered in a unipolar power equation, was arrogantly nonchalant towards this relationship though it had to face occasional moments of embarrassment, diplomatic and political.[84] Today, this mutuality resonates caution.

This coziness becomes further uncomfortable when the meteoric growth of China, in economic and technological terms is factored in.[85] China's edge has implicitly challenged US supremacy and disturbed the dynamics of power equations. Under these circumstances, the US estrangement with India has turned unproductive. In the world *realpolitik*, only India can be pitted against China to blunt its edge and cause diminution of its stature and deflect its disproportionate influence. As a result, the US strategic paradigm may have to shift to gradually wean India off Russia and to position India on the other end of the seesaw with China. But India is underweight in economy and malnourished in technology and, thus, needs to acquire commensurate ballast to politically, economically and technologically fit this role, if assigned. The Indo-US nuclear dea,l with attendant dismantling of sanctions, could yield a useful growth spurt but more may be wanted. The US may, thus, like to align its foreign policy to best serve its interests and perpetuate the American-led global order.

India and the US
India and the US have a history of decades of mistrust and commercial alienation on transfer of high-end technology. The American paranoia worked to curb India's soaring ambitions in nuclear research and led to strict

sanctions on sophisticated technology, even for peaceful purposes, because of its dual-use potential. The illusion of risk was bloated beyond reality and believed as such. This stance continued for long despite India's impeccable credentials and vociferous assertions of atoms for peace as well as doctrinal denial of first-use of nuclear weapons.[86] Understandably, the European allies also followed this lead silently and submissively. This reinforced the denial to India of commercial access to technology. The blockade was really severe.

Times have since changed and the American perception of India's missions and programmes has altered greatly. The Bush Administration made a significant shift in the foreign policy paradigm and the concomitant civilian nuclear deal of 2008 between India and the US has caused a thaw in their relations and paved the way for concerted economic interaction. As a result, many of the denial regimes and commercial sanctions have been lifted on bilateral trade. Politics and bureaucracy no longer stand in the way of scientific research and transfer of technology. Therefore, both countries can now seize this initiative, build a benign relationship and usher in a new phase of fruitful cooperation. They must also shed past inhibitions and prejudices so as to take optimal advantage of attendant opportunities and mutual strengths. The trust deficit accumulated over the years must be 'creditised'.

Fortunately for India, it is no coincidence that the budget proposals of the Obama Administration want NASA to outsource rocket development for the human space flight programme to private commercial companies.[87] Of course, the first choice would be the US enterprises like Space Exploration Technologies or United Launch Alliance, etc. yet India can offer a cost-effective proposal with the requisite delivery schedules. There are bound to be other competitors like Russia and China. Here, Russia certainly has much expertise to launch crewed spacecraft but has never landed astronauts on the moon. Besides, its own space programme is in dire straits.

Again, China too has spotted this lucrative opportunity but its hands are full due to its own carry-over of commitments and ambitious projects in the pipeline. In comparison, "India's space expertise at this point rivals China and may even exceed it in certain key scientific areas such as lunar landers and telemetry."[88] The assessment appears correct. There are, however, other contenders such as Japan and South Korea that have varied experience, but India's chances are arguably higher due to its competitive expertise, abundant trained manpower, success-quotient and value for money factor.

Conclusion

India has no proclaimed long-term national policy and its space activities have been personality-oriented or success-impelled. Nevertheless, India's commitment to derive benefits for the socio-economic uplift of the masses has remained unwavering and continues as an espoused objective. However, as a spin-off of the space applications, India has gained expertise in remote sensing and geo-spatial usages and today its latest satellite, the Cartosat-2B, launched in July 2010, is world-class and compares well with the best in the field. India needs to maintain its lead in this space specialisation and retain its niche for commercial advantage.

India has made repeated assertions at the UN and other international forums about its opposition to the arms race in outer space. This policy stance stands as the space doctrine of India to guide its research and development in space technology. Lately, however, there has been a strong and persistent demand from military commanders and security strategists to optimally harness the available space assets for passive defence applications like intelligence, surveillance and reconnaissance. This demand appears legitimate in the contemporary global security scenario and can be readily adapted from existing space resources at a minor extra cost. India is opposed to covert activities in principle and has yet to openly decide on the issue of military usages.

In the domain of space strategy, there is an urgent need to formalise a unified set of strategic goals, and against this template, to identify gaps and disconnects in policy and execution. The larger objectives can then be broken down to specific thrusts and pointers for research so that the results can be expedited and operationalised. Arguably, these would be legal uses within the *fides* of the space treaties and bearing no encumbrance of any ban under the corpus of space law. India would be well advised to legitimately take diverse advantages of the existing resources within the policy framework.

India did not get space technology on a platter and earnestly sought it from friends and foes alike. The results were mixed but India's indigenous efforts never faltered and it achieved formidable progress. Today, India is nearly self-reliant and has found a respectable berth in the array of space-faring nations. In consequence, it has embarked on a commercial trail to help other developing countries. The quality and reliability of its space products and services compare well globally, while, at the same time, the offerings are competitive and cost-effective. The space business is expected to be brisk.[89]

International ramifications of India's space policy touch China rather intimately. This is because China considers India as a rival, competitor and an adversary too. On the other hand, China is growing too fast and too big. Worst still, it is indulging in an ostentatious display of its military might in the space arena with grit and gumption. This is disturbing and has the prospects of upsetting the global power equilibrium. It is here that India can crucially help stabilise the dynamics of a unipolar global order. Alternatively, it can be achieved by the US galloping ahead to considerably enhance its lead in the economy and attain higher echelons of technology. The suitability of the option can be worked out by the US to best serve its interests.

In the end, it seems pertinent to highlight that the world has become flat and there has been the death of distance.[90] The globe has become a level playing field with equal opportunity and equal participation, *albeit* not for crass competition but fruitful collaboration. India has attained proficiency and gained experience with a high success-quotient in space technology. Therefore, India and the US can be mutually benign partners and the latter may find it lucrative to outsource mundane products and non-sensitive applications to India and reap the benefit of its cost-effective performance *vis-à-vis* quality assurance. Fortunately, trade relations between the two countries are flourishing, to mutual advantage, consequent to the Indo-US nuclear deal. Only the policy decisions at the appropriate level need to be expedited. This sounds pragmatic euphony.

Notes

1. *Space Security: Need for a Proactive Approach*, Report of the IDSA-Indian Pugwash Society Working Group on Space Security, Academic Foundation, New Delhi, 2009, p. 18.
2. Ibid.
3. Joe Katzman, "India's Emerging Military Satellite System," August 10, 2005, http://www.windsofchange.net/archives/007318.php, cited in Harsh V. Pant and Ajay Lele, "India in Space: Factors Shaping the Indian Trajectory," *Space and Defense*, Vol. 4, No. 2, Summer 2010, p. 51.
4. V. S. Mani, "Space in Ancient India" in V. S. Mani, S. Bhatt and V. Balakista Reddy, eds., *Recent Trends in International Space Law and Policy* (New Delhi: Lancer Books, 1997), pp. 22-23.
5. *Sri Guru Granth Sahib* is a compilation of psalms written by the Sikh Gurus and saints from the Hindu and Muslim religions and even belonging to the low castes. This Book is written in the Punjabi language, in the Gurmukhi script. It contains psalms written in Sanskrit, Persian, dialects of Punjabi and some languages of India like Rajasthani, Marathi, Brajbhasha, etc. It was compiled more than four centuries ago. For more details, refer G. S. Sachdeva, "Cosmos in Theological Scriptures" in *Astropolitics*, Vol. 10, Nos. 3 & 4, 2012.
6. The Thumba Equatorial Rocket Launching Station (TERLS) is located at Thiruvananthapuram in Kerala (South India). India has provided free access to all nations to launch experimental rockets from this station as a gesture of goodwill to humanity.
7. Quoted from the Second United Nations Conference on the Exploration and Peaceful Uses of Outer Space. National Papers:India, UN Document, A(Conf 101/NP/6), May 8, 1981, p.15.
8. ISRO, *Sarabhai on Space* (Bangalore, 1979), p.11. Also refer Mani, et. al., eds., n.4, p. 5.
9. Satellites are called birds in industry parlance.
10. This situation can be illustrated by the project on engine development for launch vehicles given up midway at the Aeronautical Development Establishment, Bangalore.
11. For the same financial year, the US allocated $14 billion for NASA. Figures shown are variable due to fluctuations in the exchange rates.
12. Dollar equivalents are approximates in view of changing exchange rates. Figures have been taken from the Budget Estimates for the respective financial year. For comparison, the NASA budget for the year 2010 was $ 18.7 billion.
13. Government of India, Department of Space, *Annual Report 2011-12* supporting the Budget Estimates for the Financial Year 2012-13.

14. Ibid., p.4.
15. Ibid. Words in parenthesis added for clarity.
16. Ibid..
17. With effect from November 1, 2009.
18. *The Times of India* (New Delhi), November 1, 2009.
19. Ibid.
20. The launch of the GSLV-3 took place as scheduled on April 15, 2010, but it failed within minutes of blast-off and crashed into the Indian Ocean. The failure has been analysed to a possible cause believed to be non-ignition of indigenous cryogenic engines. ISRO plans to rectify the investigated defects. *The Times of India*, April 16, 2010.
21. Government of India, Department of Space, *Annual Report 2011-12*, p. 23. It supports the Budget Estimates for the Financial Year 1996-97.
22. This project has faced a setback due to failure of the GSLV-D3 launch on April 15, 2010. With this happening, the GSAT-4, the advanced communication and navigation satellite that was expected to provide the "eye in the sky", has also crashed.
23. *The Times of India* (New Delhi), November 1, 2009.
24. n. 21, p. 6.
25. Ibid., p. 53.
26. Ibid., p. 97.
27. Comment by Cosmonaut Rakesh Sharma while addressing a meet at the Physical Research Laboratory, Ahmedabad. Interview published in the Delhi edition of *The Times of India*, March 8, 2010.
28. *The Times of India* (New Delhi), November 1, 2009.
29. n. 21, p. 49.
30. Ibid.
31. As per standards of classification, mini-satellites should weigh 10-100 kg and micro satellites, 1-10 kg.
32. n. 21, p, 25.
33. Ibid., pp. 6-7 and 10.
34. Pico-class satellites weigh between 0.1 to1.0 kg. Other satellites carried on this vehicle were the CartoSat-2B, AlSat from Algeria and two satellites, one each from the University of Toronto and Switzerland.
35. n. 21, p. 29.
36. ISRO Press Note released to IANS on March 31, 2010.
37. n. 18. As per the latest Government of India Report, this mission is planned during the 2013 timeframe. Refer n. 13, p. 68.

38. Plans are to send two Indians into space in a crewed flight. To evaluate critical technologies, a precursor module will be launched on a modified PSLV in 2013. n. 18.
39. n. 21, p. 8.
40. *The Hindu*, July 13, 2010. Also refer, *The Times of India*, July 13, 2010.
41. *The Hindu*, July 23, 2010.
42. n. 21, p. 8.
43. Ibid., p. 25-26 and 29-30.
44. The *Times of India,* June 23, 2012.
45. *SPACE TRAVEL*, Exploration and Tourism, May 27, 2010, Washington DC (SPX).
46. n. 21, p.9. Till November 2011, 473 VRCs had been set up in 22 states and Union Territories. Refer n. 21, p. 9.
47. n. 21, p. 9.
48. Group Captain Joseph Noronha (Retd.), "Life Begins at 40 (A Life Sketch of ISRO)", published in *SP's Aviation,* Issue 10, November 2009, p. 32.
49. D. S. Watrous, "Indian Space Force: A Concept for the 21st Century", *Trishul*, Vol IX, No 2, pp. 6-7.
50. India's first dedicated military satellite for naval communication and surveillance will be launched by the year-end 2010. This would provide the Indian Navy with 'wide-network-centric operations' and 'maritime domain awareness.'
51. The Army Commanders Conference in 2007 discussed *Space Vision 2020* and urged the government on this aspect.
52. A total of 22 operational satellites at the end of 2009. Based on ISRO information.
53. G. S. Sachdeva, "Space Mines: Dialectics of Legality", in *ASTROPOLITICS*, The International Journal of Space Politics and Policy, Vol. 7, No 2, May-August, 2009, pp.135-49.
54. n. 1, p.70.
55. Prof. U. R. Rao, 6th Bose-Einstein Lecture on Science, Technology and Environment delivered at IIC, New Delhi, on February 26, 1998. Quoted from G.S. Sachdeva., "India's Space Activities in the 21st Century," published in S. Bhatt and V.S. Mani, eds., *India on the Threshold of the 21st Century--Shape of Things to Come* (New Delhi: Lancer Books, 1999), p. 276.
56. Dr. K Kasturirangan, "Space: The New Business Frontier", an interview published in *Business India*, April 26-May 4, 1997, p. 53.
57. n. 21, p. 9.
58. Ibid.
59. G. S Sachdeva, *Outer Space: Security and Legal Challenges* (New Delhi: KW Publishers Pvt. Ltd, 2010), pp. 42-43.
60. n. 21, p. 96.

61. Ibid., p. 97.
62. Ibid.
63. Ibid.
64. Ibid., p. 96.
65. Emily Wax, "India's Space Ambitions Taking off", *The Washington Post*, November 4, 2009.
66. Vinod Patney, "Planning for Aerospace Dominance", in Air Cmde Jasjit Singh, ed., *Aerospace Power and India's Defence* (New Delhi: Knowledge World, 2007), pp. 312-13.
67. Report from *PLA Daily* from Beijing published in *The Times of India*, May 25, 2010.
68. Ashley Tellis, "Punching the US Military's 'Soft Ribs': China's Anti-satellite Weapon Test in Strategic Perspective," *Carnegie Endowment, Policy Brief No. 51*, June, 2007. Also refer, Ashley Tellis, "China's Military Space Strategy" published in *Survival*, Autumn 2007, pp. 41ff.
69. K.K. Nair, *SPACE- The Frontiers of Modern Defence* (New Delhi: Knowledge World, 2006), p. 132.
70. *The Times of India*, July 13, 2010.
71. Ibid. Words in parenthesis added.
72. Opinion of John M. Logsdon, Professor Emeritus of Political Science and International Affairs at Space Policy Institute of George Washington University, USA, quoted from news report of Emily Wax, n. 65.
73. News report from Beijing in *The Times of India*, August 18, 2010.
74. *The Times of India*, August 18, 2010.
75. Harsh V. Pant, The Emerging Balance of Power in the Asia-Pacific, *The Royal United Services Institute Journal*, 152 : 3 (June, 2007).
76. n. 69, p. 156.
77. *The Daily Telegraph* news report published in *The Hindu*, July 23, 2010.
78. Based on Memorandum with the US signed in 1979. This was formally abandoned in 2001 when South Korea joined the MTCR.
79. n. 69, p. 187.
80. Ibid., p. 145.
81. Ibid., p.143. ISNET is the Inter-Islamic Network on Space Sciences and Technology.
82. n. 13, p. 9.
83. Reported in *The Times of India*, August 16, 2010.
84. For example, the repeated veto by the USSR in the UN Security Council in 1971 during discussions on the Indo-Pak conflict, also called the Bangladesh War.
85. China has cruised past Japan to become the world's second largest economy behind only the US (*New York Times* news service report published in *The Times of India*,

August 17, 2010) while the ASAT shoot has demonstrated its techno-competence. China's disputed territorial claims are also leading to verbal duels with Washington.

86. The Cabinet Committee on Security reviewed operationalisation of India's Nuclear Doctrine on January 4, 2003. For a detailed discussion on the subject, refer, Dr. Manpreet Sethi, *Nuclear Strategy: India's March Towards Credible Deterrence*, (New Delhi: KW Publishers Pvt. Ltd, 2009).
87. The US Budget, 2010.
88. Saswato R. Das, "Getting India into Orbit," *The Times of India*, May 13, 2010.
89. The latest commercial venture of Antrix has been the launch of AlSat for Algeria in July 2010.
90. Expounded by Thomas L. Friedman in his book, *The World Is Flat* (USA: Penguin Books, 2005).

4

India as Vendor of Space Utilities to Developing Countries: An Example in Cooperation

Introduction

Space facilities in the service of humanity and for the welfare of mankind have developed by leaps and bounds and are proliferating globally. Some of these like broadcasting and television, mobile telephones, internet search and mailing, global positioning system for navigation of public carriers and private vehicles, to mention just a few, have become so intertwined with our lives that it is unimaginable to survive without them. Any disturbance in these facilities may cause life to come to a halt, resulting in indescribable inconvenience and chaotic disruption.

It would be relevant to visualise the impact and implications of a net shutdown for a few days. First, the thought itself is frightening for it can wreak havoc on our information systems, intelligent working, data dissemination and even the human psyche. Most computer users resort to e-mail and cloud computing for storage of data — imagine a situation wherein they could not access this crucial information or vital data for days. The frustration, though understandable, would be immense. And this is apart from what many youngsters may suffer from: pangs of deprivation and near-psychiatric withdrawal symptoms if they are unable to, or cannot, log on the net. Computer addiction is severe, serious and rampant.

Let's also consider the fallout on various systems if the net shuts down for a few days. Suspension of e-mail services alone would leave over one billion users of Gmail, Yahoo and Hotmail in the lurch, leave aside other platforms and modicums like Facebook and Twitter. Then global e-commerce that is going to touch $ 1.4 trillion by 2015, would take a severe beating. So would on-line banking and the stock market. It would be a nightmare coming true. For web-business and industries dependent on the net, a blip for a few days will have a gross impact, leading to unsustainable losses. And, above all, it would cause colossal loss of manhours worldwide. One really shudders to imagine such a predicament.

Today, we take for granted the boon of space facilities and their innumerable advantages. This, is, however, not the pinnacle of advancement in this direction and many unknown utilities are yet to unfold. It can be easily surmised that our dependence, and very survivability will hinge on the critical devices operating from space usages. The advancement of technology is elating for, and causes pride in, human intelligence, yet the spectre of utter dependence in the future sends deep shudders of helplessness and a feeling of surrender to inevitability.

Apart from daily common-use utilities briefly alluded to above, there is a wide range of other usages for business incrementals, natural resources management, economic development, social uplift, literacy campaigns and cultural exchange, again to mention some select objectives. Besides such peaceful uses, space assets can be exploited for not-so-peaceful defence applications and national security imperatives. The inventory of space uses can be considerably enlarged and yet it shall defy comprehensive listing. The developing countries would also need all or some of these facilities, though presently on a smaller scale, and their future requirements may progressively expand with societal development, the growth of economies, habitual usage and possible addiction.

India, within the group of developing countries, is a reasonably developed space-faring nation with indigenous competencies in space technology, a commendable track record of space launchings, expertise in satellite fabrication and system integration, an enviable infrastructure of ground facilities and a vast fund of technical human resources to be rightly proud of. Antrix is the corporate face of India, offering space utilities and space products for sale. And, above all, India is committed to international cooperation with bonafide altruistic motives to make available space facilities to developing brethren nations at affordable, competitive costs and impart participative training to their manpower for future self-reliance under a programme called SHARES (SHARing of Experience in Space).[1] India has provided such yeoman service in the past to many Asian, African and Latin American countries and is expected to live up to its commitment. The developing countries can definitely take advantage of this opportunity available with a member of their comity to meet their legitimate social and economic aspirations. India, thus, sets a good example in cooperation, with the gains equitous and reciprocal.

Organisations of Developing Countries

Developing countries, following traditional wisdom, have found strength in numbers and unity. Accordingly, they have formed different associations and organisations of different kinds, having different objectives with different affiliations and different structures. As a result, their political standing gains weight and their economic bargaining power gets reinforced. Thus, their benefits, even if shared, are substantial because each country is a member of several groups. To name some of these groupings, one may mention the loosely used classification of underdeveloped countries with the commonality of poverty and social backwardness. This group has lately graduated to gain the euphemism of developing countries. This is a nebulous group where entry or exit is either voluntary or based on the perception of the developed Western world or the international funding organisations that dole out development aid funds for societal uplift and economic amelioration.

Many countries have made good use of such grants to consolidate their economies and set in motion engines of growth to effectuate the multiplier effect and have, thus, risen to a category above the developing countries, though still with rampant inherent weaknesses, like China and India and Brazil. There is another category of developing countries that has survived on regular and continuous aid and benevolent doles without igniting the internal engines of development or impelling multipliers in the economy, thus, remaining stagnant. These are countries like Pakistan and its ilk.

Asian Organisations

There are several Asian groupings which are overlapping in objectives with countries embracing membership of multiple organisations across the board. For example, the South Asian Association for Regional Cooperation

(SAARC) was established in December 1985, at the initiative of Bangladesh but based on the declaration signed by the respective Foreign Ministers in New Delhi on August 2, 1983. The original members are Bangladesh, Bhutan, India, Maldives, Nepal, Pakistan, and Sri Lanka. Afghanistan has joined later in 2007, and then Mauritius. Apart from the association members, a few countries have been given Observer Status, like the US, European Union and South Korea. Some more states like Russia and Iran are desirous of Observer Membership and a couple of Observers e.g. Myanmar wish to be promoted to regular membership. SAARC is headquartered in Kathmandu. The utility of the organisation is obvious and evident.

The association is expressly conscious of the political reality that some states in the region have disturbed mutual relations and are likely to be influenced by national perspectives and prejudiced by vested interests that may interfere with the achievement of its objectives like regional cooperation "for promoting peace, stability, amity and progress in the region."[2] It has, therefore, been mutually decided that all bilateral issues are to be kept aside and only multilateral (involving many countries) issues would be discussed, without being biased by bilateral relations, and rising up to the avowed objectives of the association. Additionally, it was resolved that all decisions in the organisation would be taken unanimously and would need a quorum of all eight members. Though apparently a restrictive clause, it appears to be a wise step in the long run. It also seems conducive to the durability of the association and for the community's betterment and sustained development through participative decision-making and egalitarian cordiality.

Another good example of an Asian organisation could be the Association of Southeast Asian Nations (ASEAN). This association was formed in 1967 with five founder-members, while today it has a membership of eleven countries, with three special members, namely, China, Japan and the Republic of Korea. This broad-spectrum organisation stands for global peace, stable progress and shared prosperity. Further, taking cognisance of the geo-politics of the Asia-Pacific region, it pledges for military cooperation apart from materialising mutual business opportunities, accelerating economic growth and impelling socio-cultural development.

Other Regional Organisations

Besides Asia, there are other regional organisations of developing countries with the same or similar objectives of cooperation for mutual benefits of economic growth, reciprocal trade expansion on the bases of social affiliations and cultural affinities. In their own humble ways, they have strived to promote international peace and cooperation as well as sharing of knowledge and technology. Such organisations are spread over all of the continents and exist in multiple numbers, offering an overlap in membership and, at times, carrying incongruous views on the same subject of international concern.

There are several regional identities. We have already discussed Asia in detail. The next could be the Asia-Pacific region with the Asia-Pacific Economic Cooperation Organisation. Similarly, for the African continent there are the African Union (AU) and Afro-Asian Cooperation Organisation. For the Arab nations are the Arab League, formed in 1945, and the Organisation of Islamic Countries (OIC). The Caribbean states have formed the Caribbean Community (CARICOM) and Association of Caribbean States. Then there is the Latin American Parliament and the Union of South American Nations. This list is not comprehensive and two more deserve mention: these are the Commonwealth of Nations formed in 1931 with the British Queen as the present head and the Commonwealth of Independent States (CIS) constituted in 1991 by the former states of the USSR. They are all aspiring for cooperation, technical and economic, that could be mutually rewarding.

Development Needs of Developing Countries

Developing countries have peculiar yet distinct affinities and are bonded in geographical proximity, historical ties, common customs and cultural moorings in shared traditions with similar humane values. Except for varying political perceptions and consequent strategic compulsions, the nations possess a criss-cross of commonality of religion like between India and Nepal or India-Bangladesh-Pakistan-Afghanistan or the Arab League, languages and ethnicity like India and Pakistan or India and Sri Lanka or Pakistan and Afghanistan or nations of the African continent, and many more such permutations and combinations can be culled out and cited. Further, despite formal border barriers, a part of the population keeps easily moving across political boundaries for various purposes and some countries have a bilateral Treaty of Friendship to facilitate trans-border movement like India-Nepal and India-Bhutan or an unwritten understanding for free floatation of population in nearby areas on both sides without bothersome official formalities.

Though the developing countries are generally steeped in social backwardness and poverty, each one is at a different threshold of economic growth and industrial advancement, social parameters of development and other indices of progress. Many of the countries are landlocked. So, for amelioration, they have been dependent on aid and relief from various sources which have mostly been linked to disaster needs and rehabilitation. Hardly any succour of this nature has been invested in business areas to generate an economic multiplier effect. For example, Bhutan, Nepal, Maldives and Mauritius sustain on tourism, Bangladesh on exports, Myanmar has just started growing, Pakistan lives off aid and Afghanistan has yet to decide whether its economy will run on doles or the traditional produce of the country. Each one has its own array of options to choose from and challenges to confront.

However, rampant poverty, widespread illiteracy, escalating unemployment and burgeoning population remain the dominant and common problems of the developing countries, irrespective of their global location. Of course, poverty remains the core issue and as per the World Bank Report,[3] there are nearly 1.2 billion people living on $ 1.25 (about Rs. 65) per day. One-third of the poor population lives in India and one-third in Sub-Saharan Africa. The remainder one-third is scattered over Asia, the Arab countries, Latin America and on the periphery of Europe. No doubt, "World Development Indicators show that extreme poverty headcount rates have fallen in every developing region between 1981 and 2010" yet "this figure should serve as a rallying cry to the international community."[4] Development is acknowledged as a Human Right.

Poverty and subaltern problems of the developing countries mentioned above are more or less intertwined and individual efforts at tackling each one are regressed by the others. Thus, despite vigorous efforts at alleviation, progress is slow and the results still slower to show up. It appears a long-gestation process with traction that inches slowly towards progress. In this scenario, India, despite rampant poverty, appears a developed big brother, is presently better off economically, has broken the shackles of social backwardness and has embarked upon a policy of industrialisation, infrastructure development, business globalisation and scientific advancement. Results have accrued from such measures and are amply visible.

Thus, though India has come out of the morass and is galloping towards progress individually, it has to shoulder a traditionally vicarious responsibility, customary in the East, of carrying the developing brotherhood along, in a spirit of friendship and cooperation, for common welfare and shared prosperity. The obligation is daunting, but nevertheless surmountable with a joint family thrust, because in a broader objective, India has already pledged to inclusive development through space facilities for Third World and developing countries and has already initiated concrete steps in this direction. On the other hand, developing countries also cannot afford to postpone their prosperity and deserve to use space applications for their social and economic amelioration.

One area of multi-purpose help for common advancement comes from the space applications of proven advantages. And Indian expertise in space utilities for societal development, distance education, communications network, broadcasting and television, and allied applications is widely recognised. India's track record of rocket launches and consecutive successes is envied even by the space powers. Again, India's acumen in system integration of satellites and capability for consultancy services is competitive in technology and comparatively cost-effective. The developing countries can have the benefit of these services available within the community at bargain prices. In fact, India has already undertaken a cooperative initiative to provide certain facilities as aid on a bilateral basis and many more utilities are available commercially through the corporate organ of the Indian Space Research Organisation (ISRO) christened and styled as Antrix.

Benefits from Space Utilities

Space technology has turned out to be a boon for societal uplift and economic development by providing satellite applications and utilities for myriad advantages which can be multiplied by simple extensions and innovations. It is conceded that the space industry is prohibitively expensive for the underdeveloped countries of SAARC yet each one need not set up or invest in independent facilities. In fact, their individual needs are much smaller than modular facilities and their best option would be to lease out a quantum suitable and proportionate to their planned requirements. With experience, lease options can be widened to accommodate greater facilities for the requisite footprint. Only a broad, and not really comprehensive, listing of space utilities for social and economic progress is given in the succeeding paragraphs.

- **Broadcasting and Television:** These are multi-media facilities with diverse range and content. These have mass appeal in information dissemination and a large and extendable footprint. This audio-visual medium offers great advantage in education, entertainment, news and views and opinion formation.
- **Communications Network:** Instant communications is the need of the times, nay, an inescapable necessity, whether in the mode of landline telephones, mobiles or internet communications like e-mail and its cognates like Facebook, tweets, various types of friends' circles, professional linkages, and so on. The present generation can hardly live without these fads and facilities.
- **E-Governance:** Governments in many developing countries are progressively switching over to e-governance for prompt and accurate information to its citizen. This is not a trend but an actual necessity of the times and the demand of the public for its availability at every level, including the rural population from the villages.
- **Rural Development Package:** The economies of most of the SAARC countries are agrarian, and depend on agriculture for which the majority of the population is based in the rural areas that have their own typical problems of village living, lack of basic facilities, rural mindset and stubbornness about traditional methods of farming. This society has to

be introduced to new techniques of cultivation, new variety of seeds, more effective insecticides and weedicides and the latest fertilisers for better and improved production. Space facilities offer viable solutions for such problems.

- **E-Literacy:** Space applications can be effectively used for e-literacy to impart basic education to the masses sitting at home, at their own convenience and, free time, without the need to attend classes in schools. They can provide a virtual classroom facility with advanced teaching methods.
- **E-Health and Tele-Medicine:** Space applications as tools of delivery of health care and medical facilities were encouraged under UNISPACE-1999 as Millennium Development Goals (MDG). These have since come of age and have been extended to control of epidemics of tropical diseases: plague, malaria, and so on. Medical transcription, e-surgery and seeking of second opinion are commonplace usages of modern times. Mobile health has come to stay for its virtual benefits and life saving applications.
- **Meteorology and Weather Forecasting:** These areas have assumed great importance due to agriculture and crop regulation, in sowing or harvesting, to escape inclemency of dry spells or heavy rainfall or thunder storms and similar calamities. Information about weather has also become important to daily life for various reasons. Thus, weather forecasting, meteorological monitoring and climate related studies have assumed importance.
- **Disaster Management and Rescue Operations:** Disasters seldom come announced but may give a short warning. Therefore, the need for space facilities for disaster forecasting, finding of the location, damage control, rescue operations and overall management of the disaster site and situation is vital and inescapable.
- **Navigation Facilities:** The need for timely and accurate navigational facilitation of all types of transportation systems which are gaining in speed and proliferating in routes has become important for the safety and security of their operations. Transportation systems could be land-

based like rail and road networks; or aerial like aircraft, civil or military; or naval ships and waterborne vessels whether at sea or in inland waters.
- **Satellite Imagery for Natural Resources:** Exploration of deep-embedded natural resources was not easy, but remote sensing facilities have made such discoveries simpler. It has also become convenient to find water tables and their underground courses. And, of course, to determine the extent of forestation or deforestation poses no problem with satellite cameras. Therefore, remote sensing facilitates reconnaissance of natural resources, including vegetation, for optimum exploitation to ensure proper management of scarce resources for long-term sustainability. Thus, this provides a vital opportunity for informed and systematised decision-making based on data from earth-observation satellites.[5]

Space Leadership of India

India's space policy is qualitatively directed towards societal betterment and economic progress. Thus, "India's focus has been entirely on civilian applications for social and economic development with very little attention being paid to leveraging space assets or technologies for security or strategic planning."[6] The speech by Vikram Sarabhai on the occasion of the inauguration of the Thumba Equatorial Rocket Launching Station (TERLS) in 1968 articulated this direction with the semblance of a space policy for India which even today seems equally valid and relevant for the SAARC nations.[7] He said:

> There are some who question the relevance of space activities in a developing nation. To us, there is no ambiguity of purpose. We do not have the fantasy of competing with economically advanced nations... But we are convinced that if we are to play a meaningful role nationally and in the community of nations, we must be second to none in the application of advanced technologies to the real problems of man and society, which we find in our country.[8]

It seems pertinent to allude to another assertion by Sarabhai in his address as Scientific Chairman of the United Nations Conference on the Exploration and Peaceful Uses of Outer Space on August 14, 1968. His message reiterates the strong linkage between community development and peaceful uses of space technology. The wisdom enshrined here can serve as useful lesson and motivator for the SAARC countries in their development efforts through space technologies and their applications for peaceful purposes. India can well serve as a template for progress with portability of projects as well as dispel misgivings about the objectives by offering its experience as an example. He then said:

> I believe that several uses of outer space can be of immense benefit to developing nations wishing to advance economically and socially... It is necessary for them to develop competence in advanced technologies and to deploy them for the solution of their own particular problems, not for prestige, but based on sound technical and economic evaluation involving commitment of real resources... Indeed, they would discover that there is a totality about the process of development...[9]

In pursuance of the above policy, India operates the largest socio-economic developmental network in the world with a variety of space applications. This focus has not shifted and space-related applications continue to be an inalienable strategy for community welfare and uplift of the masses. This stance is acknowledged by the National Aeronautical Space Agency (NASA), which has applauded the country's efforts in using space missions for societal needs.[10] The applications encompass telecommunications, broadcasting, tele-medicine, tele-education, and rural development applications. Lately, with Gramsat, the scope and footprint of activities has been broadened so that "information-centric" benefits can permeate to the grassroots level through Village Resource Centres (VRC) that have been established throughout India.[11]

Complimented for these achievements, India has come of age in the space sector, and has been acknowledged not only as a space-faring nation, but also as a space power due to own prodigious standing in technological research, vehicle launch record, footprint of applications, integration of satellite systems, deep outer space probes, control and monitoring of satellite operations, remote sensing capabilities, infrastructure facilities and allied expertise in space activities. India has emerged as a space power, and is accordingly reviewing its achievements in space-related applications, assessing its peaceful initiatives in geo-spatial data acquisition, realigning its objectives based on new technology developments, and redefining its strategies in consonance with the geo-political realities. The SAARC nations can, therefore, take advantage of India's high points and seek consultancies on country-related aspects.

Space is a profitable industry and affords business prospects in activities relating to uses of space, such as telecommunications, direct television broadcasting, remote sensing, and internet technologies that have led to a global space market for satellites, launch vehicles, and earth-based ground facilities. India has gained expertise and experience in satellite fabrication, systems integration, software development, launch vehicles, developmental applications, remote sensing, and data acquisition facilities. Particularly in the areas of remote sensing and geo-spatial monitoring, Indian satellite facilities rank equal to the global best and are well sought after; India has carved its own niche of specialisation. Therefore, the commercial space policy in India has started offering on commercial basis space hardware, launch facilities, software applications, and consultancy services through Antrix, the commercial arm of ISRO, established in 1992. Commercial products and services from India are relatively cost-effective and high-end on state-of-the-art technology.[12] Further, the Indian experience in small satellite development and low-cost launch facilities makes for a bargain offer.

With regard to earth-observation, India has developed and operated advanced high-resolution remote sensing systems in the IRS and CARTOSAT series of satellites. Further, India has pioneered in the launch of multiple small satellites. To illustrate, the PSLV-C9, which also launched CARTOSAT-2A, established a world record of launching nine small satellites comprising the indigenous IMS-1, and eight small satellites for international customers in a single launch. And PSLV-C15, which launched CARTOSAT-2B in July

2010, also put in orbit three small satellites, including one from India called StudSat that was developed by university students,[13] and two additional ones from Algeria and Switzerland.[14]

Another feather in India's cap is that through April 2011, the PSLV project has recorded 16 consecutive launch successes proving high-reliability and soundness of design of the Indian launch vehicle. And India is expected to launch the first satellite of the Indian Regional Navigational Satellite System (IRNSS) (a constellation of 7 satellites) aboard the PSLV C-22 in June 2013. This satellite will provide position accuracy of better than 10 m and shall extend its reach by 1,500 km around the country. It is designed to provide accurate real time Position, Navigation and Time (PNT) services to users on a variety of platforms with 24x7 service availability under all-weather conditions," Thus, the IRNSS shall provide two basic services: a standard positioning service for common civilian users and a restricted service for special authorised users. The neighbouring SAARC countries can usefully plan to take advantage of this venture because of its substantial overreach.[15]

Competitive Alternative

The requirements of developing nations relating to space utilities can be obtained from any of the space powers like the US, Russia, the European Union or China but they may not unbundle their package for such small, individual and customised needs. Besides, their costing may be much higher due to the inputs of high labour costs, possibly beyond the affordability of these states. The US has not ventured into offering commercial transactions to the outside world and also its costing may be relatively high. The Russians have started offering space products and services like space trips for private individuals and crew transportation for the US. but it may be wary of entertaining small customers. The Chinese quotes, however, could be much lower but with a caveat that there may be strings attached and the seller country may be selective in choosing partners, on a common ideology or for strategic advantage, for bilateral relations. Thus, there may be a dilemma for some developing nations.

Other noticeable Asian suppliers could be India, Japan, South Korea, and, to a limited extent, Pakistan. For instance, Japan has a well-developed and multifaceted space programme that is credible and visible. Today, Japan enjoys dual-use capabilities of its space assets and can boast of Command, Control, Communications, Computers, Intelligence, Surveillance, and Reconnaissance (C4ISR) capabilities; space-based communications, navigation, positioning, and targeting; and a Ballistic Missile Defence (BMD) architecture. In due course, Japan may commence use of space assets for security and military ends, and may soon project itself "as a significant military force to contend with, not only in the Asian continent, but in the entire world."[16]

Notwithstanding, Japan's space probes have been primarily fixated on a scientific orientation that concentrates on planetary explorations and discoveries in astro-physical sciences. This is borne out by Japan's seven-year sojourn to an asteroid by Hayabusa[17] and its planned launch to Mercury in 2014, using a craft that will be covered in mirrors to reflect the heat of the Sun, while, at the same time, designed to be partly powered by solar energy.[18] Thus, the current commitment will leave little opportunity for Japan to be a vendor of space utilities.

South Korea is a much later entrant to the club of space-faring nations, and it has developed its space capabilities with US technology transfers.[19] It has tried to build its expertise on this baseline, but "to make up for lost time, South Korea has adopted a mid-entry strategic approach...to leapfrog to the requisite technology levels through technology transfers." This strategy was coupled with a "smart selection" approach to "select areas of development based upon its unique needs and resources, instead of trying to accomplish overall efficiency in all areas."[20] It is a reasonable guess that South Korea will have neither intent nor spare capacity to lease out customised facilities in small measures to developing countries.

Pakistan offers no real competition to India in the space field. In fact, Pakistan's space programme comprises more turnkey projects proliferated from China with considerable dependence and little indigenous component. Reports on Pakistan are replete with authoritative references to covert transfer of technology by China and minor technology chasms bridged with the help of the United States. It appears that Pakistan has not assimilated space technology and has always looked for "quick-fix" solutions from trusted allies and friends. "It can be safely surmised that while Pakistan has immense ambitions, both military and civil, with regard to space, it currently lacks the wherewithal for the same."[21] Thus, it is in no position to reliably help the developing countries to realise their aspirations.

India, on the other hand, has a reliable fund of tried out facilities, is cooperative in attitude and cost-effective in space commerce. India has started from scratch in the development of space technology. It never enjoyed the luxury of transfer of state-of-the-art space technology from the developed Western space programmes, such as of the United States and Europe. Since India was challenged by many commercial embargoes and technology denial regimes during the Cold War, the country relied on creative improvisation, indigenous research, and reverse engineering to develop its space programmes and the bottom-line is that it has permeated the projects at the shop-floor. The results are there for all to see in India's single-handed achievements in the space arena.

It is important to note that India has developed the requisite technology, system integration, and fabrication capabilities on its own, and today, many Indian-built satellites have been successfully orbited with indigenous launch vehicles. Communication and remote-sensing satellites have been in operation for long and continue to be further improved. These satellites are contemporary in technology and rank high in performance on global benchmarks. Their reliable and sustained functioning in the space environment stands testimony to the technical excellence achieved in India.[22] And, of late, success has become institutionalised with ISRO. India has, therefore, crossed the threshold of space technology from a developing status to an advanced developed space programme.

Moreover, India has mustered an adequate pool of qualified and research-oriented scholarship for technology development in the space area. This manpower is motivated and dedicated to breach technology barriers and is equally poised to make many a significant breakthrough. Even today, India has a vast resource fund of suitable manpower for the tasks of innovation as well as production.[23] In view of the delineated status, India's technology standing, experience in the field, cost-effectiveness of its space products and, above all, its willingness to vend customised facilities, India appears an attractive choice for the developing nations.

Occasional Insensitivity of India Towards Developing Countries

India itself was once a developing country and fully realises the compulsions, handicaps, helplessness and frustrations of this league while treading the path of development. So it can fully empathise with their plight, particularly its neighbouring countries in South Asia like Afghanistan, Bangladesh, Bhutan, Maldives, Nepal, Sri Lanka and others. One thing common to these countries is that they are all developing countries and most can ill-afford to independently invest in space facilities. Their initial experimental projections fit in for leasing or sharing of space facilities. Many have, rightly and wisely, resorted to this option. But India has, in a few cases, such as Bangladesh, Bhutan, Maldives and Sri Lanka, insensitively spurned their requests, for commercial help in initialising space facilities for the reason of economic viability of the project. India also failed to offer alternatives or make available consultancy to evolve other viable options suitable to their respective needs.

As a result, these countries, disillusioned and disappointed, have resorted to other sources to meet their demand. China has been proactive in the field with goodies on offer and they have turned to China to seek solutions and help. This turn of events has caused genuine concern to India. The restricted vision of techno-consideration by ISRO may be strictly professional but is neither best suited to India's strategic interests nor in consonance with the committed foreign policy of friendly relations with neighbours. This slant needs correction to a holistic analysis of such international public tenders. The narrow focus of evaluation needs to be commensurately broadened and that too immediately.

The Case of Nepal

Recent news emanating from Nepal through Xinhua, the Chinese news agency[24] is rather disconcerting. Nepal had been allotted an orbital slot for a satellite in 1984. This slot, if not utilised before 2015, would lapse back and claim to it would be lost. Though rather late in the day, Nepal has set up a feasibility study committee to explore the technical possibility and source of procurement to launch its first broadcasting and weather forecast satellite within the target of validity. For the present, to facilitate TV broadcasts and weather forecasts, Nepal's TV channels and weather forecasting offices are paying around US$ 25 million a year for accessing international satellite services. Nepal is also conscious that to begin with, it may not be able to fully utilise its capacity for internal consumption and may have to commercially lease out the spare capacity, possibly to India or China or both.

Meanwhile, experts in Nepal have stressed the need to develop the whole process of satellite launching through a joint venture of national and international firms with coordination by the Government of Nepal. The project would require a huge amount of investment and of necessity, it would be desirable to adopt the public-private-partnership model. As Nepal finally goes ahead with long overdue plans to launch its first satellite before 2015, the country may turn to China, that has, in the recent years, helped a number of developing countries, including some of India's neighbours like Bangladesh and Sri Lanka, with technological and financial assistance for their satellite programmes. In fact, China's Great Wall Industry Corporation (GWIC) has helped launch satellites for a number of developing countries ranging from Pakistan and Sri Lanka to Bolivia and Nigeria. And now China can be expected to pitch in, in a determined bid for cooperation with Nepal for this venture on enticing terms. It seems to have already sounded Nepali diplomatic circles on this aspect.[25]

Other Such Cases

Another word of caution. ISRO, through its corporate entity of Antrix, has also cold shouldered a few international tender notices or has defaulted in positively responding to specific queries from SAARC countries like Bangladesh, Bhutan, Myanmar and Maldives[26] terming such projects as technically or economically unviable. The internal analysis by the organisation may be objective and valid yet the countries could be appraised of feasible alternatives or advised for consultancy to suitably modify the projects. Inaction in such cases can have a political fallout of serious dimensions, detrimental to national interests. Corporations like Antrix must superimpose global reality and strategic national interests over the canvass of economic considerations to make country-specific responses with a holistic approach. However, we need not feel disconsolate over a few misappreciations as long as long-term lessons are learnt and imbibed.

The Lessons

China, of course, is active, in fact, proactive, in the South Asian region and has proffered proposals with dual offers of technological assistance and financial support through soft loans on favourable terms from the China Development Bank. These offers have found favour with many developing countries. Thus, China's recent success in launching satellites, particularly for countries in India's neighbourhood, has concerned New Delhi. In consequence, there was a meeting with officials from different government ministries in April 2013 to come up with a strategy to respond to China's uncomfortable moves. In fact, the Cabinet Committee on Security, earlier, in March 2013, had asked ISRO to become more responsive and active in responding to neighbours' needs.[27] This advice was the result of reports from the Research & Analysis Wing (R&AW) that highlighted how India's lack of interest in such requests and tender notices in the recent past had encouraged China to continue its spree of success in this field. Continuance of adherence to such stark commercial considerations will be detrimental to India's strategic interests and may even impact India's global standing.

The above advice to ISRO also stems from the political compulsions of SAARC Charter which mandates reciprocity of cooperation for mutual benefits. Further, India, as a matter of policy, is committed to international cooperation with bonafide altruistic motives to make available space facilities to developing brethren nations at affordable, competitive costs and impart participative training to their manpower for future self-reliance under a programme called SHARES (SHARing of Experience in Space). The insensitive attitude of ISRO and Antrix to the commercial enquiries from neighbouring countries, however small in value or unviable in technological concept, runs counter to India's policy commitment to the principle of cooperation. Thus, if not the product offer, at least a consultancy offer can be made. Hence, proper amends need to be made and promptly too.

It needs to be appreciated that in such a cooperative environment, the cold economic logic of commercial viability does not reign supreme and commends reasonable compromises tempered with country-specific relational realities. Political sensitivities and cultural affiliations must be duly acknowledged and equally respected in the interest of stable cohesion, organisational traction and durable unity of the region. Antrix must adopt a crafted policy of wider consultations before refusal or rejection of similar offers. Such action will embellish its image as a corporation of strategic importance with responsible commerce as an instrument of deft diplomacy. Further, if deemed necessary, ISRO/Antrix may consider positioning a resident strategy expert or a Foreign Office representative on deputation.

Conclusion

In a nutshell, the world has got used to, nay, addicted to, the use of facilities provided from space assets, in their daily personal lives and professional work. It seems near impossible to live normally without their constant operational availability. The dependence on these facilities has grown as subservience and usage have become inevitable. Developing countries, too, have started appreciating the spectrum of advantages in relation to alleviation of poverty, illiteracy and allied societal ills and are aspiring to share this phenomenon on their own terms. Of course, these countries cannot afford to individually invest in such expensive space facilities yet do not want to be left out of the mainstream. A viable alternative is to share or lease their limited requirements from service providers. Among all the dependable service providers, India would certainly prove an operationally reliable, economically competitive and trustworthy partner.

India is one such country that has competence and capacity and willingness to offer such facilities and consultancy services, with no strings attached, commensurate to the threshold of contemporaneous needs. India is part of the region and can offer state-of-the-art services as a local vendor with cost benefits. The developing countries can surely reap the benefits of India's offer. India's economics makes sense: it is a tried and tested regional member, a dependable friend, with no hegemonistic ambitions. And this offer neatly fits into the objectives of regional cooperation mandates. Hence, India can prove to be comparably the best service provider of space products and facilities, which are reliable in performance and economical in costs, to the developing countries in South Asia, even Africa and Latin America. India can now confidently sell space dreams and make them come true, at the same time, though it must be appreciated that the burden of expectations can be exceedingly high and demanding. It is truly so, but India can carry it bravely and magnificently in the spirit of avowed cooperation.

Notes
1. G. S. Sachdeva, *Outer Space: Security and Legal Challenges* (New Delhi: Knowledge World, 2010), pp. 42-43.
2. Preamble to the SAARC Charter.
3. Some details published in *The Times of India* (Chandigarh Edition), April 19, 2013.
4. Ibid.
5. Also refer G. S. Sachdeva, "India's Space Activities in the 21st Century", in S. Bhatt and V. S. Mani, eds., *India on the Threshold of the 21st Century—Shape of Things to Come* (Lancer Books, 1999), pp. 257-288.
6. *Space Security Need for a Proactive Approach*, Report of the IDSA-Indian Pugwash Society Working Group on Space Security, Academic Foundation, New Delhi, 2009, p. 18.
7. The Thumba Equatorial Rocket Launching Station is located at Thiruvananthapuram in Kerala (South India). India has provided free access to all nations to launch experimental rockets from this launch station.
8. The Second United Nations Conference on the Exploration and Peaceful Uses of Outer Space, National Papers of India, UN Document, A (Conf 101/NP/6), May 8, 1981, p. 15.
9. Indian Space Research Organisation (ISRO), *Sarabhai on Space* (Bangalore: India, 1979), p. 11. Also refer to V. S. Mani, S. Bhat and V. B. Reddy, eds., *Recent Trends in International Space Law and Policy* (New Delhi: Lancer Books, 1997), p. 5.
10. "Exploration and Tourism," *Space Travel* (Washington, DC), May 27, 2010..
11. Government of India, Department of Space, *Annual Report 2011-12*, p.9. "Already more than 473 Village Resource Centres in 22 states and Union Territories" have been established in India.
12. Also refer G. S. Sachdeva, "Space Policy and Strategy of India" in Eligar Sadeh, ed., *Space Strategy in the 21st Century* (Routledge, 2013), p. 312.

13. *The Hindu*, July 23, 2010..
14. Ibid. Also, refer *The Times of India*, July 13, 2010.
15. http://www.indianexpress.com/news/india-to-launch-first-navigational-satellite-in-june/1089086/0#.UURgJe-kB_4.gmail. Accessed on March 17, 2013.
16. K.K. Nair, *SPACE: The Frontiers of Modern Defence* (New Delhi: Knowledge World, 2006), p. 156.
17. The probe returned with samples from the asteroid in 2010. NASA plans a similar mission to return with samples by 2023. News Report from NASA of May 26, 2011, as published in *The Hindu*, May 27, 2011..
18. *The Daily Telegraph* news report published in *The Hindu*, July 23, 2010.
19. Based on Memorandum of Understanding with the United States signed in 1979. This was formally abandoned in 2001 when South Korea joined the Missile Technology Control Regime (MTCR).
20. Harsh V. Pant, "The Emerging Balance of Power in the Asia-Pacific", *The Royal United Services Institute Journal*, 152: 3, June 2007, p. 187.
21. Ibid., p. 145.
22. For a detailed analysis, refer Sachdeva, n. 12, pp. 303-321.
23. Of course, India has missed an opportunity to participate in the International Space Station programme, which could have yielded an advantage of trained and experienced astronauts.
24. Press Release of May 5, 2013. Also reported in *The Global Times*, May 6, 2013.
25. Ibid.
26. Maldives has still kept the window open for India, despite the closure date. It is understood that negotiations at governmental level are expected to take place in India in April 2013.
27. *The Hindu*, May 6, 2013.

5

International Cooperation As Core Concept of Space Law: For Diplomacy and Confidence

Introduction

Cooperation is instinctive in living species and coexistence their intuitive nature.[1] This principle pervades human institutions, international law and, at the same time, is the core concept of space law. Even in inter-state relations, international cooperation has been the *sine qua non* in all state activities, and yet in practice, international cooperation is the first to be sacrificed in the pursuit of national interests or conflict situations. On terra firma, man has somehow managed to survive, though, at times, under difficult predicaments or near the precipice, but outer space is one realm where risks *sans* cooperation are bound to be ultra-hazardous and even self-annihilatory. The choice, though difficult, is in our hands and should be made with utmost wisdom.

Frankly, breaching of outer space has been quite a game changer for international law, as has been decolonisation in the 20th century. The comparison is strikingly similar and the results near-analogous. As a consequence of entry into outer space, new subjects have emerged, new concepts have emerged, new doctrines have emerged, new dogmas have emerged, new ethics has emerged and new jurisprudence has emerged, and so on. Changes are evident, distinct and perceptible. The ushering in of new ideas, new thought-processes and new mindsets for the nascent space law sets it singularly apart from the traditions of international law. This has been a significant break and a conceptual transition away from international law that preached coexistence, yet space law that advocates cooperation has steadfastly retained its maternal bonds with international law. Nevertheless, a vacuum occurred in the legal order of outer space, though temporarily, till it was partially filled by the Outer Space Treaty.[2]

The evolutionary process undergone by international law finds a certain degree of parallelism, if not similarity, with that of the regime of space law albeit compressed in time. What really happened to international law over centuries has caused space law to mature in decades. In fact, space law initially grew as an adjunct of international law but metabolised faster. It, however, did not sever its moorings from international law. But now, it is getting metamorphosed into an independent and auto-poietic system[3] with nexus and cross-linkages to other sub-systems of cognate legal regimes,[4] international jurisprudence, objects and subjects of international law, their operations and other multiple applications.

The intrinsic complexity of space law, thus, becomes discernible as its contours gradually crystallise its new formations and its grammar becomes more distinct and communicable. However, there is no reductionist approach of criminality in space law; the bottom-line is either common survival or collective annihilation. The stakes are clear and the choices really limited. The decision, however, is of our volition and depends upon our feeling of altruism and level of sagacity. International cooperation, therefore, is the mandate of space law addressed to the international comity for adherence irrespective of their being space-farers, space-users or space-watchers.

From another angle, legal pedagogues of the West and protagonists of traditional theories have, for centuries, vehemently maintained and consistently taught that international law, from medieval times, has been the law of civilised nations, for civilised nations, and operated between civilised nations, thereby implying the European states or Western Christian civilisation,[5] that practised civilised conduct in international relations and, thus, their conduct and behaviour could be regulated by mutual pacts and relatively soft laws. "Practically all the European writers endorse or support this view."[6]

This view, by corollary, becomes readily adaptable to space law, the corpus of which comprises some UN General Assembly Resolutions, a couple of treaties, a few agreements and a lot of guidelines and in-principle statements. A scrutiny of this documented law brings to clear focus a basic tenet of international cooperation in space activities. An attempt is made in this chapter to vindicate this hypothesis by examples and illustrations from space law and concomitant state practice which will highlight and confirm that by mutual cooperation, nations of the international comity would be able to usefully interact and fruitfully operate without fear, let or hindrance in outer space and on celestial bodies to reap the intended benefits.

Thus, international law as the law of civilised nations can be confidently emulated by drawing a striking parallel to assert that space law is the law of cooperative nations. It, thus, makes a discernible revelation that while international law attempts to regulate relations among a society of states, space law transforms governance into a legal order of a genuine state community.[7] Whether overawed by the ultra-hazardous nature of space activities or the intuitive knowledge of the inherent shortcomings of technology or the ultimate inadequacy of own capabilities, the need for cooperation comes up in sharp relief, with emphasis on equality and mutuality. The difference may appear subtle yet the impact of this orientation is profound and positive.

In fact, the exploration and use of outer space and the celestial bodies is unthinkable without intensive and diversified international cooperation in contemporary foreign affairs as well as in regulating specific relations between states arising from joint scientific and technological activity. It can, therefore, be proclaimed that international cooperation in outer space is an unconditional obligation of all countries, whether space-faring or so aspiring. Perhaps, we could move a decisive step forward and assert that "cooperation in space matters is a duty recognized by all states."[8] This opinion has been endorsed by the statement that "…the observance of the principle of cooperation is an obligation of states."[9]

The Nature of Space Law

From time primordial, outer space has been governed and regulated by the rule of God, the laws of nature and the principles of astro-physics. In fact, the necessity of international law arose only recently to regulate human activity and operation of human-made objects in outer space and the use of outer space. It is axiomatic that the primary purpose of law on the planet earth, is to ensure peace and order by informing of normative and lawful behaviour and regulating human activity within the precincts of good conduct. A Spanish proverb succinctly sums up this wisdom. It states that it is not the fence that protects the orchard but the fear that goes with it and this element of fear is imparted and instilled by the law. If the intended purpose is effectuated, at least largely, then the law has substantially achieved its objective and, in consequence, safety and security and public order seem reasonably assured.

This principle can be conveniently extended to outer space where this *specialibus* law performs a vital function by its proclamations of permitted, prohibited and regulated activities. It, thus, defines the nature of activities in outer space and stipulates the norms of human conduct in relation to this medium. The purpose of the law is prevention of aberrations by informing of good behaviour, and in the worse case, sanctions on offenders. Hence, the law seems to foster a modicum of public order in outer space. So far, this specific law has, largely, succeeded in achieving this task within the outer space and from outside.[10] As a result, no major dispute or conflict has arisen or remained totally unresolved.

There is no denying that space law mostly comprises soft law, apart from a couple of treaties and a few binding agreements. A serious study of this subject uplifts one with a sense of altruism reflected in the concept of cooperation among citizens and between countries, as enshrined in this law, to cajole its subjects into obedience. The space law can, for certain, usher in an era of peace and development through cooperation and assure fulfillment of the common interest of humanity and improved quality of life of man on the earth, in outer space and on celestial habitats.

In principle, every human activity in every field needs a law, not to curb freedom but to regulate its operation and to make it compatible with similar rights of others. It implies innocuous governance, and not invasive restraint, in the best interest of all, because rights have correlatives in duties and both must coexist for the benefit of all, that is provided for all. It also seems relevant to mention that culture is the manifestation of humanity's soft power and it becomes crystal clear that core values which cultures may individually cherish generally surround the core of cooperation. This concept, in turn, gets extended to international level to constitute and consolidate the fundamental premise of space law.

Space Law Prior to OST

The pre-OST era was the period of true and genuine cooperation by the states when treaty law did not exist and the UN had assumed the responsibility and authority to regulate activities in outer space under Article 1 (4) of the UN Charter which states that the UN shall be a centre for harmonising the actions of nations in the attainment of common ends. In pursuance to this mandate, the General Assembly took upon itself under Article 13 of the Charter to initiate studies "encouraging the progressive development of international law and its codification"[11] relating to outer space. The UN General Assembly became active and ensured the passing of resolutions that established the principles of non-appropriation in outer space and on the celestial bodies, freedom of exploration, extension of international law and applicability of the Charter of the UN as well as the Statute of the International Court of Justice to regulate human activities and artificial objects in outer space.[12] As part of its efforts in this direction, the General Assembly, within the framework of the UN, constituted an international body for cooperation in the study of activities in outer space that is now called the Committee on the Peaceful Uses of Outer Space, in short, COPUOS.[13] The importance of cooperation among states is boldly underscored here.

The next important stage in the evolution of space law dawned with the General Assembly, adopting by acclamation[14], "The Declaration of Legal Principles Governing the Activities of States in the Exploration and Use of Outer Space."[15] The declaration enunciated cardinal principles which later got embodied in the Outer Space Treaty[16] of 1967. The fundamental principles underlying the declaration were, and continue to be, cooperation, assistance and consultations among the states. These principles find copious references and due emphasis in the total gamut of space law spread over treaty law, agreements, guidelines and principles promulgated by the UN.

The Mandate of International Cooperation

Civilised nations do not have to be mandated to act and foster harmony in their respective and common best interests. It behoves nations to naturally behave in that manner and, of course, we all claim to be civilised nations honouring the tenets of international law and cooperating in international relations. This mandate assumes greater significance in the extraordinary medium of outer space where our dependence becomes all the more pertinent and national support systems too tenuous to infuse enough confidence in operations up there. Thus, cooperation tends to be obligatory *inter se*, mutually supportive and confidence building in such an environment. The benefits are bound to be extensive and reciprocal.

No wonder, a strong thread of cooperation links through and intrinsically binds treaties, agreements, guidelines and principles germane to space law. International cooperation, thus, becomes the fulcrum of space law to stabilise the growth of international, global, political, economic and scientific development. Further, to act as an enabler, the United Nations has, indeed, assumed *suo motu*, responsibility as a focal point for international cooperation in outer space, and the nodal agency designated for the purpose is the Office of Outer Space Affairs (OOSA) at Geneva. The UN also provides support to this viewpoint from its own Charter which purposefully pledges, "To achieve international cooperation in solving international problems of an economic, social, cultural, or humanitarian character…"[17] And it needs no casuistry to extend this principle to space matters.

Subsequently, the postulate of state cooperation became one of the fundamental principles of international law, which were unanimously confirmed by all UN member-states in the Declaration on Principles of International Law Concerning Friendly Relations and Cooperation Among States[18] in accordance with the UN Charter. The declaration, thus, ordains "…cooperation among states regardless of their political, economic and social systems, in different areas of international affairs…[as] an international legal obligation."[19] Its binding nature is, thus, apparent to be harnessed to maintain international peace and security in accordance with the principles of sovereignty, equality and non-interference. This amply demonstrates a sense of compulsive cooperation among nations. But to convincingly prove this hypothesis, we need to illustrate with a few examples.

Mandate for Cooperation in OST

The OST is replete with references to cooperation between states and the relevant provisions are meaningful and well-intentioned. To begin with, the Preamble to the treaty optimistically exhorts states-parties, "...desiring to contribute to broad international cooperation in the scientific as well as the legal aspects of the exploration and use of outer space for peaceful purposes." It further believes "...that such cooperation will contribute to the development of mutual understanding and to the strengthening of friendly relations between states and peoples."[20] The hope is sincere and the attitude realistic towards global amity and world peace.

Outer space "shall be the province of all mankind"[21] shifts the emphasis from the traditional postulate of national sovereignty to international cooperation with community rights for the common good, thus, highlighting the underlying principle that there are areas where the common interests of mankind must be served jointly and given primacy. This clause concedes the possibility of a conflict of ideology or a clash of national interests in space operations but dispels "any such spectre to seek a common vision of their future relations in a newly accessible environment."[22] This principle strengthens the sense of the international community with *de facto* respect for other countries and encourages common security for the sake of mankind. The principle enunciated must be safeguarded zealously even under the pressure of vested national interests.

It is, thus, held by many scholars that space law contains stronger cooperative duties than general international law. Rudiger Wolfrum has particularly stressed that this principle marks a significant break and a conceptual transition away from the traditional international law of coexistence to a new law of cooperation.[23] Similarly, Rudolf Dolzer holds that the structure of space law is based on active cooperation and mutual assistance, complemented by specialised duties towards activities in and relating to outer space. This reflects the concept of obligation of assistance with voluntary spontaneity and in the form of reciprocity.[24]

Article I of the treaty, while referring to the freedom of scientific investigation in outer space and celestial bodies mentions that, "...States shall facilitate and encourage international cooperation in such investigation." Again Article III, while permitting space activities urges for "...maintaining international peace and security and promoting international cooperation and understanding." Further, Article IX determines that states-parties "...

shall be guided by the principle of cooperation and mutual assistance." Article X endorses international cooperation to afford "...an opportunity to observe the flight of space objects launched by ... States." Extending the principle, Article XI commands to "promote international cooperation in the peaceful exploration and use of outer space...including the Moon and other celestial bodies,...to inform...of the nature, conduct, locations and results of such activities."[25]

Further, astronauts have been exalted in status and accorded the privilege of being envoys of mankind in outer space with the intent to involve all states into this notional concept and elicit their unflinching cooperation for supporting space activities.[26] This provision entails certain obligations on states to render all possible assistance in the event of accident, distress or emergency landing, and avoid any danger to the life or health of the astronauts. In effect, the very basis of the Agreement on the Rescue of Astronauts, the Return of Astronauts and the Return of Objects Launched into Outer Space, 1968, is essentially to invoke cooperation, and, as stated in the Preamble, it has been "prompted by sentiments of humanity." The primacy of human values has been upheld and the tenor of the emphasis is clear and explicit.

This aspect of the ultra-hazardous nature of activities in space and the human propensity to solicit cooperation stems from our frailties. Quantum physics predicts risk probabilities and possibilities of failure of mechanical processes in space endeavours. This analytical data superimposed with past failures has often punctured myths about our technological arrogance and claims of zero-tolerance in accidents. Empirical observation belies our beliefs and prods us to solicit outside succour. This realisation instills a paranoia of self-inadequacy and dependence that, in turn, originates a sense of interdependence and conditioned cooptation. The chance of failure and attendant risks magnifies, and become haunting, thus, prompting mutual cooperation and reciprocity in the offer of facilities. The rationale for cooperation now becomes evidently clear.[27]

In the case of weaponisation of outer space, the OST goes a step farther than mere cooperation and binds states-parties to "undertake not to place in orbit around the Earth any objects carrying nuclear weapons and any other kind of weapons of mass destruction, install such weapons on celestial bodies, or station such weapons in outer space in any other manner."[28] The treaty permits activities "exclusively for peaceful purposes." Nonetheless, the present scenario reveals an unregulated and asymmetrical race in defence strategies and space-specific weapon technologies. The trust deficit at this threshold is risky and any mistake, advertent or inadvertent, may leave no second opportunity. The problem evades a technical solution and the usage of weapons is prompted by a mix of political, military and strategic considerations rather than academic logic or value-based decision. There could, at times, be a crisis of conscience with consequences of dangerous proportions. Hence, cooperation, with confidence-building measures, is the call of the treaty. This would assure our mutual safety and reasonable security in space and from space.

Another aspect of cooperation gets highlighted by Article XI of the OST relating to sharing of scientific knowledge from space activities with maximum publicity. The treaty reiterates that state-parties "agree to inform the Secretary General of the United Nations as well as the public and the international scientific community, to the greatest extent feasible and practicable, of the nature, conduct, location and results of such activities" or any unusual occurrence or phenomenon. This would effectuate the emphatic hope of the Preamble to promote the development of mutual understanding and the strengthening of friendly relations between states and peoples. The fundamental right of everyone to receive relevant information is deeply embedded in this clause, and cooperation is widely solicited to implement the provision. In a way, this rightly ushers in scientific socialism.

The Moon Treaty
The Moon Treaty[29] too, in its very Preamble, exhorts "to promote on the basis of equality the further development of cooperation among States in the exploration and use of the Moon and other celestial bodies." Again, in Article 2, it maintains, "All activities on the Moon…shall be carried out…in the interest of maintaining international peace and security and promoting international cooperation and mutual understanding and with due regard to

the corresponding interests of all other States-Parties." The tone, it will be noticed, is conciliatory and cooperative.

Article 4 directs, "Exploration and use of the Moon shall be the province of all mankind and shall be carried out for the benefit and in the interest of all countries... to promote higher standards of living..."[30] It further adds, "States-Parties shall be guided by the principle of cooperation and mutual assistance in all their activities concerning exploration and use of the moon. International cooperation in pursuance to this Agreement should be as wide as possible."[31]

The Moon Treaty reiterates some of the fundamental principles of cooperation enshrined in the OST. For example, Article 5 emphasises the necessity of utmost publicity and dissemination of information to the public and international scientific community of the missions launched and their results. Again, Article 6 encourages furtherance of scientific investigation, sharing of samples of minerals and other substances obtained from the celestial bodies and the desirability of exchanging scientific and other personnel on expeditions to, or installations on, the moon to the greatest extent feasible and practicable.

Guidelines and Principles

The corpus of space law, besides treaties and agreements, comprises several guidelines and in-principle documents and codes of conduct regulating space activities and governing its uses. These are contained in resolutions adopted by the UN General Assembly and are based on appeals to maintain public order in the respective field of activity and urge for voluntary cooperation to best satisfy individual needs and simultaneously reap optimum benefits for mankind.

A significant example could be the principles for regulating activities in the field of international direct television broadcasting by satellites.[32] This area involves allotment of slots for geo-synchronous satellites and sharing of the telecommunication spectrum and radio regulation. Here, the International Telecommunication Union, established in 2002, has played a sterling role in stimulating cooperation for amicable settlement of frequency allotments and adherence to the same. Granting that under the international space law, every state has an equal right to conduct legal activities in this field, directly or by entities under its jurisdiction, it appears imperative that it "should be carried out in a manner compatible with the sovereign rights of States, including the principle of non-intervention as well as the right of everyone..."[33]

Hence, "These activities should accordingly be carried out in a manner compatible with the development of mutual understanding and the strengthening of friendly relations and cooperation among all States..."[34] Further, paragraphs B, H and I of the document also encourage the same principle. In fact, paragraph D is fully devoted to international cooperation that "should be the subject of appropriate arrangements." In the same spirit, it urges "[s]pecial consideration...to the needs of the developing countries... for the purpose of accelerating their national development."[35] An altruistic attitude is clearly recommended.

Another similar document enunciating international cooperation is the Principles Relating to Remote Sensing of the Earth from Space[36] which covers a very sensitive issue so delicately enshrined in the corpus of space law. The provisions germane to the subject are sketchy and scattered in different instruments yet the principles are unequivocal in urging for international cooperation. This comes in sharp relief in the paragraphs stating Principles V, VIII and XIII. To amplify, Principle V states that "remote sensing activities shall promote international cooperation ...for participation therein... [that] shall be based...on equitable and mutually acceptable terms."[37] The exhortation to cooperation and consideration of the sensitivities of others is clearly reflected in this principle.

The UN Principles Relevant to the Use of Nuclear Power Sources in Outer Space[38] is another illustration of cooperation among states. This document explicitly "[r]ecognizes that for some missions in outer space, nuclear power sources are particularly suited...due to their compactness, long life and other attributes." But their radioactive substances carry the "risk of accidental exposure of the public to harmful radiation..." This requires "a thorough safety assessment, including probabilistic risk analysis..." for radiological protection and nuclear safety.[39] It seems highly relevant to allude to Principle I that requires the activities to be in concordance with the tenets of international law as well as the stipulations of space law. This imposes a duty and obligation to ensure the safety of the public and, thus, cooperate with the law. Of necessity, this would prompt consultations,[40] as reasonably practical, with other states and, as a corollary, to provide prompt assistance to states that are in need of this.[41] The scenario in outer space is analogous to that on the earth. Even the problems are similar because there is no technical solution to the concomitant risks.

Another pertinent example is that of UN Guidelines for Mitigation of Space Debris, 2007, that lament the quantum of existing accumulation of detritus, particularly in the lower reaches of outer space and the almost exponential addition to the same every decade. In remedy thereof, it specifically solicits cooperation among states and coordination of technological resources for reduction in redundant satellites, scavenging of space litter, avoidance of incidental debris and a conscious effort to shun advertent creation of junk in outer space.[42] The guidelines would ensure that space operations are safe, viable and sustainable into the distant future.

Code of Conduct for Outer Space Activities
Another earnest effort at global cooperation in space activities emanates from the European Union Code of Conduct for Outer Space Activities[43] (CoC) that encourages cooperation, consultation and mutual assistance to "seek solutions based on an equitable balance of interests." This accepts to redefine and impart a new understanding of sovereignty that is different from the one imbedded in traditional international law. The EU has urged nations worldwide to join and adhere to this code to make its acceptance almost universal. The CoC was to be opened for signatures by all countries in 2012 but has now been postponed till a later date. However, there is a school of thought that doubts the utility of such an exercise in repetition, and touts its futility. But this argument can be countered by citing the success of the Hague Code of Conduct against Ballistic Missiles Proliferation (H-COC) that has been signed by 148 countries.

The CoC expresses apprehensive concerns on the ever increasing space debris to highlight the risks for safety of space operations and the need for mitigation for avoidance of collision hazards. It also alludes to the growing weaponisation of outer space and seeks codification of practices for transparency and confidence-building measures without disturbing the inherent right of states for collective self-defence under the provisions of the UN Charter. It also exhorts for adherence to the essential treaties like the OST, the Rescue Agreement and the Convention on International Telecommunications and Radio Regulations. Lastly, it urges members to make their national space policy and procedures in tandem to minimise accidents, and ensure that solutions to disputes are based on an "equitable balance of interests."

Frankly, there is nothing new in this CoC that is not already contained in the treaties, various agreements and many UN proclaimed principles and guidelines on different issues. The reiterations make it an apparently innocuous document with undefined vocabulary but it is this utter innocence that makes it suspect. The UN Guidelines for Mitigation of Space Debris, 2007 is not very different in content and spirit or at least in its implications on the specific subject. And its urging for responsible behaviour seems a mere platitude. Despite this, there are remonstrations to the obvious that have been stated in the CoC.

The Chinese objection relates to the requirement for open declaration of the space policy of states. This is hardly serious enough to merit reconsideration. On the other hand, India laments the lack of an institutionalised enforcement mechanism and legally binding provision for verifications. This is a sensitive issue and impinges on sovereignty. It is surmisable that in its turn, India may also be wary of such overt openness and revealing exposure to verificatory and intrusive inspections. And further, both countries hold the grouse that they were not part of the creative process and have not contributed their perspective. Whatever be the background to offering or denying this opportunity, the reservations seems trivial and stem purely from the ego of their geo-political weight and diplomatic formality.

In the US, the Stimson Centre of Washington University has prepared a draft of a Model Code of Conduct (MCoC) that is being keenly debated. At the same time, NASA is also working independently on a similar Code of Conduct for Outer Space though it has come up with no major differences with the CoC. In all probability, it may not be substantially different from the European Code or the Model Code except that this would reflect the US perspective and be expressive of its overt consent. Its appeal to the world at large is uncertain till promulgated but its boldness, if at all, would be limited to the US self-interest and continued hegemony. This is understandable unless it turns otherwise objective and altruistic to become an international instrument of wide solicitation and durable value.

Some of the Asian nations (like China, India, Japan and South Korea) have also come of age in space affairs and should participate in such formulations proactively and express their viewpoint eloquently to protect their genuine interests rather than just acquiesce into them belatedly. The European effort appears incomplete in scope and repeats the obvious; and, thus, falls short of its own objectives. It needs to be comprehensive and should not restrict itself to enunciating solutions to contemporary concerns. It should be futuristic and proactive to encompass other ensuing issues like proliferating nuclear devices and power plants in outer space or elaboration of rules relating to commercial space mining, the common heritage of mankind and other matters that clamour for urgent attention.

It is from this mark that Indian and Asian space-farers can ambitiously embark to draft a framework called the Asian Code of Conduct that should be a norm-shaper. This proposal can take-off following a different approach and perspective to address common imperatives while still retaining its popular appeal and universal acceptance. It may be suggested that this initiative must also involve Australia in a genuine dialogue to gain better clout and attention. From the Asian prioritisation of major issues concerning outer space, the escalating arms race, and not space debris, assumes greater importance. Whatever be their common views regarding holistic security, space traffic management or modalities of global governance or reservations about publication of space policy or enshrining of rules and procedures in national legislations, they should find the correct iteration and right place in their draft document. This would be the right attitude rather than staying aloof on egotist credentials.[44]

Generic Variants of the Mandate

Apart from direct references to international cooperation, the space treaties, agreements as well as UN declarations and guidelines have used other generic variants for this mandate though with nearly the same connotation and implications. The term international cooperation may appear simple yet there are many layers to the concept and many levels of meaning to fully realise its nuances. Some of the commonly used terms are mutual assistance and coordination; provide technical assistance and help; necessary consultation among states; prompt dissemination of information; voluntary sharing of information; reporting of unusual occurrences; mutual consultations on hazardous experiments; assisting in identification of space objects; strengthening of friendly relations between states and visits; inspection on stations and facilities on a reciprocal basis or on request; and verification of compliances.

Such and similar expressions of an advisory nature or as pseudo-obligations are copiously found in the OST, Moon Treaty, Agreement on the Rescue of Astronauts[45] and Convention on Registration of Objects Launched into Outer Space, 1975 Further, the relevant principles and guidelines[46] too cast a specific obligation on consultation for direct satellite broadcasting. Here, the terminology or jargon used being different is not of importance but emphasis needs to be placed on substantive intent and objective purpose that clearly points to international cooperation which should be willingly imbibed and sincerely practised in space relations without any form of harmful interference in the activities of other states.

An example could be discussed. Mutual consultation is a recommended mode of cooperation contained in almost all the treaties relating to outer space and the celestial bodies. This term is new to legal lexicons; but it has gained popularity and serves as a non-legal tool of diplomacy to legal obligation. It provides for periodic meetings to discuss common interests and examine measures to promote the objectives. The term consultation comprises methods for prevention or resolution of disputes or acts as for consensus-building on issues of common concern or interest. It, thus, provides a reasonable opportunity for clarifying the factual situation to arrive at a mutually agreed upon solution. Its forms and functions are varied and its nature could be obligatory or optional, written or oral. All the same, "consulatations" as a mechanism are gradually acquiring both legal significance and effectuation to usher in harmony and peace.

One can also solicit support from the views of Goedhuis that in meeting the varied challenges of the space age, man has been able to combine the forces of the social complex which makes us realise the world's interdependence because of the limitations of technology and the ultra-hazardous domain of outer space that necessitates a cooperative social organisation in consequence thereof. With these dominant values and international organisations, peaceful and shared use, rather than ownership or exclusive control, has become the primary theme. The inclusivity of all states, technically capable or still struggling, is inalienable and integral to the order of outer space. Thus, an important feature of space law reflects the gradually transforming structure and a process to detoxify international relations and recognition of the compulsion of international cooperation in the field of outer space.[47]

UN Declaration on International Cooperation

Further, not to miss the importance of cooperation as a fundamental principle of space law, it seems highly pertinent to refer to the Declaration on International Cooperation[48] adopted by the UN General Assembly and solely devoted to promoting and fostering the theme of international cooperation among states on an equitable and mutually acceptable basis. This declaration was adopted based on the report of the Committee on the Peaceful Uses of Outer Space (COPUOS).[49] The declaration "recognises the growing significance of international cooperation among states and between states and international organisations..." and strongly advocates "further strengthening...in order to reach a broad and efficient collaboration in this field for the mutual benefit and in the interest of all parties involved...taking into particular account the needs of developing countries,"

The declaration places special responsibility on "countries with more advanced space capabilities" to offer technical assistance as requested and project an appropriate conduct to enable "developing countries ...for reaching their development goals." In the end, it exhorts, "All states should be encouraged to contribute to the United Nations Programme on Space Applications and to other initiatives in the field of international cooperation in accordance with their space capabilities and their participation in the exploration and use of outer space." The intention is clear and the obligation still clearer.

It needs to be clarified for good understanding that cooperation and competition create the template of relationships between nations yet these are not antithetical nor adversarial concepts excluding or negating each other. Friedman's law of cooperation and social development supports this opinion.[50] In fact, cooperation and competition can, in practice, amicably coexist on principles of equity, fairness and good conduct at the political as well as economic level. Hence, in contrast to competition, cooperation is not retrograde to the progress and development of mankind. On the contrary, it can work as an enabler in tandem and a catalyst to growth by constructively eliminating impediments in relational paradigms.

Encouragement by the United Nations

In international relations, independence is equal; dependence is mutual and obligations reciprocal where it is not a question of choosing a right ally but being the righteous partner. In such a predicament, cooperation is the keyword and watchword in space law and becomes a metaphor for peace, harmony and development of the world and humanity. It needs to be propagated through the efforts of the United Nations, international organisations and nationally by the governments of the states. The United Nations, thus, has an important role to play in avoiding crass exploitation while encouraging cooperation among the fraternity of nations and fostering peace on the globe and in the cosmos. It needs to involve itself in moderating, restraining and even censuring the aberrant conduct of nations in outer space to assure sustainability of the environment. This reflects a need for changing patterns of cooperation with the ultimate motive of durable and lasting peace.

At the organisational level also, the UN may consider upgrading the Office for Outer Space Affairs (OOSA) to a full-fledged and comprehensive world space organisation (or authority) for an integrated holistic approach to the various peaceful uses of space and coordinate the activities of users and dedicated organisations so as to enable everyone to reap the advantages of synergy and simultaneously avoid wasteful duplication. Such an institutional structure seems to be a prerequisite to effective space exploration and use in the interest of mankind. This calls for redefinition of its responsibilities for better coordination and discharge of onerous duties on reporting of activities and procedural dissemination thereof. Moreover, it should be adequately empowered to effectively handle sensitive space matters, to police infringements, cater to futuristic contingencies, legislative initiatives and inter-country disputes. Therefore, it must negotiate global governance where shared values of cooperation and harmony should prevail for the human good and welfare of humanity. That is the *noblesse oblige* of the United Nations.

In fact, the states should equally actively encourage and adopt international cooperation as part of state doctrines and accept this principle as an object and purpose of national legislation. It would, thus, facilitate effective implementation of national policies and procedures to ensure prompt communication of the appropriate information to the concerned international bodies as required under international law. This would involve a shift in attitude from the politics of grievances to the politics of aspirations, acknowledging that human dignity transcends national frontiers and cannot be captive to geography. Sincere and active participation by the states is vital and ultimate on this issue.

As a result of the above efforts, one can optimistically surmise that international cooperation in outer space will soon transform itself into a universal and fundamental principle to get elevated as *jus cogens*[51] of space law. It will, thus, hopefully, assume the status of a peremptory norm of state behaviour and human conduct in outer space for universal obedience and strict compliance, where aberrations will attract collective censure and sanctions. Such *jus cogens* will, thus, be able to draw substantive support from the Vienna Convention.[52]

Empirical Reality of Statecraft

Human history is witness to a grand empirical fact that at the end of each and every war on the globe, all parties, including the victor, have invariably come to the negotiating table for a final solution in a compromise over the reason for the war. President Obama has said that instruments of war do have a role in preserving peace but no matter how justified war may be, it only promises human tragedy.[53] If we can perform this amicable act after the occurrence of devastation and death, why can we not cooperate and negotiate before the catastrophe? Sagacity lies in choosing the right option, at the right time, rather than regretting the holocaust *post facto*. Peace cannot be held hostage to egotism, imprudence or irresponsibility. And space law has wisely chosen this option of international cooperation as a mandate to obviate disputes, settle differences and, finally, to negotiate and resolve conflicts. The espoused alternative is relational redressal and settlement of disagreements through consultation and cooperation among states. The wisdom of prudent governance is obvious.

The era of the space age started in the late Fifties of the last century. The launch of the Sputnik-I on October 4, 1957, by Soviet Russia was the first foray into outer space, realising the dream and vision of humanity. It altered humanity's perception of itself and its place in the cosmos. The pride and prestige was evident in the achievement and the Soviets hailed it as symbolic of, and in the interest of, universal peace. It was evident that the Soviets did not fully realise the significance of their technological achievement and its impact on world politics or Cold War relations. There was little propaganda to extol this event or extract any mileage for the benefit of foreign policy or political purposes. However, Sputnik-II followed shortly after, with spectacular success and a propaganda blitz. On this occasion, full advantage of the propaganda was taken by the Communist leadership and the committed media "to encourage incipient Soviet nationalism."[54] It implied total domination because according to the US perception "success in space implied superiority on earth."[55]

Be that as it may, the US certainly was taken by surprise and the mass media whipped up a national security paranoia while the US was, in fact, ready with the technology, infrastructure, space systems, reconnaissance satellites and other wherewithal. The irony was that the US dithered in decision-making. It fixed no target with immediacy for the launch and was diffident about the Soviet and international reaction to violation of air space sovereignty as also, it dwelt for too long on moral considerations of using the Nazi-origin V-2 rocket that was already tried and tested. By now, success had been claimed and the US arrogance on technological leadership had been badly dented.[56] The US had to do something different, a better marvel to redeem its laurels. The next best thing was landing on the moon and this was achieved first by the US.

An early impression of the uses of outer space pointed towards military ambitions and hazardous explorations, making it appear a realm of conflict and danger under the spell of the Cold War. However, the empirical reality has been different. The three decades from 1960-90 turned out to be an era of fruitful cooperation and détente in space activities despite the brinkmanship of the Cold War that was at its peak. This period was characterised by international cooperation and mutual support that gave an impetus to the development of joint space endeavours and evolved state practices that had enduring effect and efficacious compliance. This attitude can be demonstrated by state practice, superpower cooperation, the work of international organisations and ease in negotiation of multilateral agreements. A culture of cooperation was in the making. A few examples are cited in support of this contention.

The Period of Superpower Détente
First, for space forays, national sovereignty in air space never came in conflict though it could have been violated at blast-off or reentry into the atmosphere. The acquiescence was voluntary and universal. In fact, the United States was apprehensive about a contrary reaction, particularly from Soviet Russia. But the Soviet lead unwittingly and unintentionally solved this dilemma for the US. The world at large was amazed at the feat and no country raised any objections. It, thus, came to be established that space vehicles would not infringe state sovereignty in the international air space. By inference, it confirmed that national sovereignty did not operate in outer space that was free to all for peaceful activities.

Another example of superpower cooperation relates to the signing of the Partial Nuclear Test Ban Treaty in 1963 in the period following the Cuban Missile Crisis. In fact, the US had conducted nuclear tests in the upper atmospheric regions to determine the effects. The experiment demonstrated that nuclear radiation had a tendency to fall back on the earth due to gravity. The outcome could have deleterious effects on the earth and earthlings and had to be avoided in future. Negotiations ensued and a treaty was readily finalised. It was a harbinger of détente.

The restored parity in space related capabilities and surmised equalisation of the strategic weapons arsenal infused the confidence to enter into a limited rapprochement. As a result, to thaw the Cold War relations, talks on strategic issues like arms limitation were initiated. It augured well and SALT-I (of 1972) and SALT-II were successfully concluded. The arms control agreements were highly significant even though with only symbolic value. This was the first sign of self-confidence and mutual trust between the superpowers, recognition of equilibrium, and willingness to codify their equated relationship as great powers. Another similar instance was the Seabed Treaty of 1971 that dramatised their improving relations to the global audience.

A dramatic symbol of the new desire to constructively cooperate was Henry Kissinger's idea of a "hand-shake in space." Under the Apollo-Soyuz Test Project (ASTP)[57], the two countries committed themselves to scientific cooperation for utilitarian benefits in a joint manned mission where their spacecraft would link up and dock in space. This project obliged the two states to cooperate closely for three years to enable scientists to understand the working and operation of each other's systems with greater transparency and access to information. It also sought to jointly train astronauts and cosmonauts for the hand-shake mission with the notion of space cooperation to create a milieu of a new political reality. The space link-up took place on July 17, 1975.[58] Pravda had then philosophically declared, "Earth is the planet of mankind. Cooperation in space paves the road to peace, mutual understanding and good of all the people."

Another pertinent gesture of symbolic cooperation came from the USSR. The Soviets had decided to launch Salyut-7 as part of the Salyut space station series in February 1986. At the last minute, it was renamed Mir (Myr) meaning peace in Russian language. Gorbachev's decision to rename the space station smacked of propaganda, yet it sought to reduce the obsessive focus on the superpower relationship and was designed to suggest the peaceful nature of the Soviet programme as also obliquely emphasise the American effort to militarise and weaponise space through the Strategic Defence Initiative (SDI). The challenge, though mute and symbolic, was yet, significant and discernible.

Over the years, the enthusiasm for cooperation petered out and relations got stabilised to a *quid pro quo* of national interests and strife for supremacy. There were, though, disequilibrating factors and mutual relations got adversely affected yet the ebbs and flows got evened out to sustain a fairly harmonious keel. In consequence, negotiation of treaties and conventions got relegated but enunciation of soft law in the form of principles and guidelines continued unabated albeit with slow progress.

A recent example of understanding and goodwill displayed relates to the incidence of the collision between a Soviet rogue satellite, Kosmos 2251, and a functional communication satellite, Iridium, operated by a US corporate entity. This incident occurred on February 10, 2009, at an orbit of 790 km above the earth. It had enough ingredients with the potential for escalation and confrontation of a space threat. It could reach a dangerous threshold but constraint and cooperation exercised by both countries let the incident pass without demur despite claimable financial losses and concomitant political embarrassment. Maturity and spirit of cooperation clearly reflect in their behaviour of a shared vision and a fusion of the thought process.

Superpower Cooperation with Other Countries

The United States

During this period, the two superpowers cooperated, interacted and helped not only trusted allies but other struggling countries as well with technological help and scientific knowhow. NASA was the nodal agency for all aid and cooperation from the US. The consortium of the European Space Agency and some European countries were major beneficiaries because technology transfer to these recipients was with the waiver of restrictive regimes and licensing obstacles. Hence, they progressed fast and unencumbered in the space race.

In January 1984, President Reagan in his State of the Union Address invited friends and allies to participate in the development and use of a permanently manned space station and to share the benefits of this project. This proposal was formalised through an Inter-Governmental Agreement (IGA) signed on September 29, 1988. Twelve states, namely, the United States, Canada, Japan and nine states, the then member-states of the European Space Agency, joined this consortium and the space station was christened FREEDOM. The legal regime governing this international cooperative venture was stipulated in the agreement itself.

Many other countries across the globe and international organisations have also received material help from the US through NASA but these end users had to prove their bonafides for peaceful purposes and their credentials as authorised users. Convincing the US was not always easy. Nevertheless, many countries reaped the benefits of such cooperative overtures. India too enjoyed its share of the bounty of technical material under the Amendment of 1975 to the US Space Act. For instance, the Indian Space Research Organisation (ISRO) obtained the ATS-6 communication satellite for its Satellite Instructional Television Experiment (SITE) in 1975 from NASA. Such help was extended to 27 other countries under the aegis of the US Agency for International Development (AID). Earth remote-sensing LANDSAT data was also made available to interested countries on payment of a limited fee.

The USSR

In similar overtures, the USSR also helped committed Communist countries, Warsaw Pact nations and other socialist allies, and offered mutually advantageous help as per necessity and request. This aid and cooperation emanated from international treaty obligations in good faith and in deference to its Constitution.[59] The state address in connection with the first flight of man into outer space in 1961 hailed it as a victory of all mankind and happily placed this achievement "at the service of all nations, for the sake of the progress happiness and well-being of all the people in the world."[60]

An important example of multilateral cooperation was the *Interkosmos* scientific and technical programme that involved Bulgaria, Hungary, the

German Democratic Republic, Cuba, Mongolia, Poland, Romania and Czechoslovakia in the spirit of traditional friendship. Interkosmos was not set up as an international organisation but functioned through a legal mechanism of a council to supervise mutually beneficial programmes of concurrent scientific observation. The emphasis of cooperation was in five directions—the study of the physical properties of outer space, space meteorology, space biology and medicine, space communications, and remote sensing of the earth. Apart from the above, Communist China has been a bounteous beneficiary under various programmes and has made good use of this material and technological help. As a result, today, it comfortably occupies a berth of equal technological status with the US and the Soviet Russia.

India has also been accorded special treatment and privilege among the developing countries. The legal basis of Soviet-India cooperation in outer space was established by an Inter-governmental Agreement on Further Development of Economic and Trade Cooperation of November 29, 1973, that respected mutual sovereignty, territorial integrity, non-interference in internal affairs, equality and mutual benefit. The agreement contains a general obligation of Soviet assistance to develop and strengthen scientific and technical cooperation with no exchange of monetary funds. India's first few satellites used Russian rockets and other facilities. The USSR permitted an Indian astronaut to orbit the earth in a joint manned mission in 1984. India has also received occasional technological consultancy vital to its ongoing projects.

Cooperation in Negotiation of Treaties and Agreements
The United Nations has been a key player in the encouragement of international cooperation in the space segment and has, thus, played a sterling role in the regulation of space activity since the dawn of the space age. Its first tangible step in this direction was the constitution of the UN Committee on the Peaceful Uses of Outer Space (COPUOS) in 1961. This committee has done commendable service in debating and propagating some novel and unconventional principles applicable to outer space. It has been central to the drafting, formulation and facilitation of international

treaties which form the basis of the international law of outer space. The committee continues in its unfinished task.

The entire world has cooperated and displayed singular determination to support in unison the UN resolutions regulating human ventures in outer space. These resolutions lay down the principles governing activities in space, some of which are at wide variance with the traditional international law, yet the acceptance has been near unanimous. This followed a period of great cooperation during which many treaties, agreement and guidelines were negotiated with ease and speedily concluded. It was the season of evolution of space law in treaties and agreements. A soft law in the form of principles and guidelines was also adopted by the UN General Assembly in the later years.

A few illustrations will vindicate the above statement. Apart from drafting the UN resolutions, the most significant work of COPUOS has been the Outer Space Treaty[61] This required great deliberation because outer space being an extraordinary medium in many respects, demanded unique legal principles that would facilitate international relations and govern human activities in outer space. This treaty, thus, established the crucial understanding that outer space and the celestial bodies are not subject to the territoriality principle of national appropriation and sovereignty; there is freedom of activity in space without frontiers; states are responsible for their own activities and those of their nationals; and states have an obligation not to harm the outer space environment. These were new concepts contrary to established precepts of international law but their acceptance was without resistance or any objection.

Other treaties and agreements facilitated by COPUOS that deserve mention are: Agreement on the Rescue of Astronauts, the Return of Astronauts and the Return of Objects Launched into Outer Space. This was adopted on December 19, 1967, and came into force on December 3, 1968. It admits of national jurisdiction over astronauts and space objects; the Convention on International Liability for Damage Caused by Space Objects was concluded on November 29, 1971 and became effective from September 1, 1972; the Convention on Registration of Objects Launched into Outer Space was adopted on November 12, 1974, and entered into force on September 15, 1976. Another major instrument was the Agreement Governing the Activities of States on the Moon and other Celestial Bodies that was adopted on December 5, 1979, and came into force with effect from July 11, 1984. It still cannot boast of many adherents.

Apart from the above, a lot of soft law principles and guidelines have been recommended by the United Nations through General Assembly Resolutions. These may not, thus, appear mandatory for compliance yet their binding force is rather high in operational and functional conditions. State practice accords these high credence and adherence. The very nature of these instruments highlights the importance of international cooperation and the emphasis put on these by the states volitionally confirms the same. These are listed below:

- Principles Governing the Use by States of Artificial Earth Satellites for International Direct Television Broadcasting, 1982.
- Principles Regarding Remote Sensing of Earth from Outer Space, 1986.
- Principles Relevant to the Use of Nuclear Power Sources in Outer Space, 1992.
- Declaration on International Cooperation in the Exploration and Use of Outer Space for the Benefit and in the Interest of All States, Taking into Particular Account the Needs of Developing Countries, 1996.
- Guidelines for Mitigation of Debris in Outer Space, 2007.

Cooperation Through International Institutions

Apart from political symbolism, cooperation has also been elicited by the very technical necessities of space operations and functionalities. Such cooperation perforce has been channelised through international institutions that have discharged excellent regulatory responsibilities and coordinative work in the space domain. The inter-governmental organisations that deserve mention are: International Telecommunication Satellite Organisation, International Telecommunication Union, International Maritime Satellite Organisation and World Meteorological Organisation (WMO).

Consequent to the efforts of these organisations, it has been possible to make effective utilisation of communication frequencies used by satellites and the allocation of slotted positions in the geo-stationery earth orbit. The

WMO has done still better to preside over a system of voluntary and free interchange of meteorological data — both terrestrial and satellite — derived between the member countries. Cooperation of this kind has allowed a more rational and regulated exploitation of earth orbit-enabled sensible use of scarce radio spectrum and exploration of the solar system. The regulatory mechanisms have assured operational efficacy by avoiding overlaps and user conflicts.

The International Space Station (ISS) is another important institution demonstrating international cooperation on principles of equality and mutuality. The space station is a $100 billion project and consists of five modules. The three main capsules are: one for control consoles, one as living quarters for the astronaut crew, and one as a research laboratory. It operates with the participation of 15 nations, e.g., the US, the erstwhile USSR, China, Japan, and European Union, etc. in different aspects of research and training. India missed out on this option. Such joint ventures, though initially symbolic, eventually become instruments for the promotion of détente and international cooperation.

The outstanding feature of the ISS set-up is that it is a federative entity, displaying multinational cooperation in many ways and of the highest quality because space exploration and scientific investigation are essentially conjoint and cumulative activities with complex relationships determined by several factors. Thus, outer space forays have "been a vehicle for the creation of complex epistemic communities and international organizations."[62] In actual practice, however, cooperation may encounter problems and obstacles that cause tension and recrimination rather than being reciprocally complimentary. Another significant difficulty in the establishment and implementation of long-term joint space projects is the changes in national policy imperatives or shifts in economic priorities that may intervene to disrupt schedules and commitments yet existential cooperation, nevertheless, remains an outstanding feature.

Conclusion

It is to be expected that academics and pedagogues with other mindsets would not easily be able to reconcile with this school of thought with idealist leanings. Though some differences in approach are inevitable, their arguments may arise from mental reflexes or stem from doctrinal fixations. Their logic is likely to be slender on merit and their opinions may smack of vested interests and may bear only the force of the clout. But in the event that they prevail, peace would have lost its constituency, conscience would have been defeated and humanity would have failed in its brief. Consequently, the human race may risk reaching the precipice. The alternative is survival with dignity through existential cooperation. Let the majesty of the law triumph and prudence win for now the choice is ours, the decision is ours.

It can now be safely concluded, in a nutshell, that the architecture of space law is supported on the foundation of cooperation between nations and consultation among the states. This, incidentally, is the *sine quo non* of reciprocal relations under the guarantee of freedom and peace in outer space. Thus, just as international law is held out to be the law of civilised nations, in the same vein, space law is the law of responsible and cooperative nations that value their freedom of operation, understand their stakes of disruption and uphold good sense to appreciate their debt to future generations. In fact, with such widespread emphasis on international cooperation, there seems to be little scope for conflict and antagonist activities. Hence, in this scenario, space security appears automatically and reasonably assured, while peace and goodwill can be expected to prevail in the cosmos.

From the foregoing analysis and perusal of space history, it gets amply substantiated that international cooperation is the cornerstone of space law for the betterment of mankind and improvement in the quality of life on the earth. It is not a peripheral issue but a strong strand that runs through the paradigm and becomes the common denominator of treaties and agreements, principles and guidelines. Therefore, this concept has been fully imbibed and internalised by this nascent branch of international law and has, thus, become integral to its functioning and the normative behaviour of all states. No wonder, the aberrations have been few and minor, whereas states have demonstrated judicious restraint to eschew escalation of conflict. Under the circumstances, in this era of space civilisation, we need to look beyond security; we need to think of peace and sustainable development.

In consequence, fulfillment of the principle of cooperation and other cooperative objectives embodied in the corpus of space law, that is mostly soft law, appear potentially achievable for the first time. This extols the general philosophy of cooperative pacifism enshrined in the OST and other cognate international instruments. In the end, let me quote from the prophetic thought of the English philosopher, Bertrand Russell, "The only thing that will redeem mankind is cooperation." Wisdom lies in scrupulously following this adage.

To conclude, the binding and obligatory nature of the mandate is indisputable and seems to have acquired a certain customary force. There is also a growing peace constituency in a divided international polity that subscribes to the principle of international cooperation enshrined in space law. Logically, adherence can be reasonably expected, mostly voluntarily or at times through sanctions. Empirically, this has happened, the ordained cooperation has been mutual and the requisite restraint elicited voluntarily.[63] The hypothesis is vindicated. *Quod Erat Demonstrandum.* And now, not to foster and propagate this concept would be unforgivable.

Notes

1. Observation of insect and animal behaviour reveals the existence and evolution of democratic institutions of equality and shared group decisions. They also exhibit a high degree of cooperation and a spirit of sacrifice in the interest of the group or community.
2. The Treaty on Principles Governing the Activities of States in the Exploration and Use of Outer Space, including the Moon and Other Celestial Bodies, 1967 (in short, OST).
3. Anthony D'Amato, "International Law as an Autopoietic System", in Rudiger Wolfrum and Volker Robens, eds., *Developments of International Law in Treaty Making* (Berlin, 2005), pp. 335-399.
4. For example, analogous regimes of Antarctica and the high seas and aspects of international law like state responsibility, state liability *et al*.
5. For views on the contribution of the Asian states to the evolution of international law, refer R. P. Anand, *Development of Modern International Law and India* (New Delhi, 2006), p. 2.
6. Ibid. Also Charles G. Fenwick, *International Law* (Bombay, Second Reprint 1967), p 56. Several writers with this viewpoint are quoted in Ibid., pp.6-7.

7. Detlev Wolter, "Common Security in Outer Space and International Law", UN Institute of Disarmament Research, Geneva, 2005, p.111. *UNIDIR/2005/29.*
8. S. Bhatt, "International Problems Concerning Use of Space," *International Studies, JNU*, Vol. 12, No. 2, 1973, p. 269.
9. Y.M. Kolosov, "International Cooperation in Mass Media", *Sovietskoye Gosudarstvoi Pravo*, No. 2,1978, p. 32.
10. This is a personal statement and assessment. Many scholars may express varying shades of disagreement.
11. Ogunsola O. Ogunbanwo, *International Law and Outer Space Activities* (The Hague: Martinus Nijhoff, 1975), p. 11.
12. General Assembly Resolutions 1148 (xii) of 1957 and 1721 (xvi) of December 20, 1961.
13. General Assembly Resolution 1348 (xiii) of 1958.
14. Adopted on December 13, 1963. Ogunbanwo, n. 11, p. 14.
15. General Assembly Resolution 1962 (xviii) of 1962.
16. The Treaty on Principles Governing the Activities of States in the Exploration and Use of Outer Space, including the Moon and Other Celestial Bodies, 1967. (In short, OST.)
17. *The Charter of the United Nations*, Article I, paragraph 3.
18. *UN General Assembly Resolution* 2625 (XXV) of October 24, 1970.
19. V. S. Vereshchetin, *The Principle of Cooperation in International Space Law and its Implementation in the Soviet Union* (Moscow: Science Publishing House, 1977), p. 211.
20. The Treaty on Principles Governing the Activities of States in the Exploration and Use of Outer Space, including the Moon and Other Celestial Bodies, 1967. The Preambular paragraphs.
21. Ibid., Article I, para 1.
22. Wolter, n. 7, p.85.
23. Ibid., p. 251. Rudiger Wolfrum, "The Problems of Limitation and Prohibition of Military Uses of Outer Space", *ZaoRV*, Vol 44, 1984. Also refer R. Wolfrum, "Common Heritage of Mankind", in *Encyclopaedia of Public International Law*, Vol 11, 1989, p.67.
24. R. Dolzer, "International Cooperation in Outer Space", *ZaöRV*, Vol. 45, 1985, p. 527. Also refer n. 16, p. 92.
25. n. 16.
26. Ibid., Article V. For a detailed analysis, refer G. S. Sachdeva, "Astronauts as Envoys of Mankind in Outer Space: Resolution of a Dilemma", *Asian Journal of Air and Space Law*. 1 AJASL, 2011, pp. 1-22.

27. This doctrine seems to seek inspiration from the Buddhist philosophy contained in *Pratiya samut pada* which emphasises the inadequacy of sentient beings and preaches the *dharma* of cooperation.
28. n. 16. Article IV.
29. Agreement Governing the Activities of States on the Moon and other Celestial Bodies, 1979.
30. Ibid., Article 4 (1).
31. Ibid., Article 4 (2).
32. Principles Governing the Use by States of Artificial Satellites for International Direct Television Broadcasting, 1982. UN General Assembly Resolution 37/92 (annex) adopted on December 10, 1982.
33. Ibid., Paragraph A—Purposes and Objectives of the Document.
34. Ibid.
35. Ibid., Paragraph D—International Cooperation.
36. UN General Assembly Resolution 41/65, adopted on December 3,1986.
37. Ibid.
38. UN General Assembly Resolution 47/68, adopted on December 14, 1992.
39. Preamble to the Resolution.
40. Ibid., Principle 6.
41. Ibid., Principle 7.
42. Like the testing of ASAT missiles in outer space by China and the US in 2007 and 2008 respectively.
43. The European Union Code of Conduct for Space Activity, October, 2010. (In short CoC.) Refer Michael Listner, "An Update on the Proposed European Code of Conduct," *The Space Review,* August, 8, 2011.
44. For more details on the subject, refer, Rajeswari Pillai Rajagopalan, "Debate on Space Code of Conduct: An Indian Perspective," *ORF Occasional Paper # 26,* October 2011, Observer Research Foundation, New Delhi.
45. Agreement on the Rescue of Astronauts, the Return of Astronauts and the Return of Objects Launched into Outer Space, 1968.
46. n. 32.
47. D. Goedhuis, "An Evaluation of the Leading Principles of the Outer Space Treaty of January 27, 1967," *NTIR,* Vol 15, 1968, p. 40. Also refer Fenwick at n. 5 *supra,* p. 110.
48. Declaration on International Cooperation in the Exploration and Use of Outer Space for the Benefit and in the Interest of all States Taking into Particular Account the Needs of Developing Countries, 1996. UN Document Supplement No. 20 (A/51/20).
49. A/CONF.101/10 and Corrigenda 1 and 2.

50. Expounded by Thomas L. Friedman in his book, *The World Is Flat* (Penguin Books, 2005).
51. *Jus cogens* are customary laws that reach close to *jus naturale* to be deemed implied agreements that are obligatory in nature and require no specific ratification. Benjamin N Cardozo, *The Nature of Judicial Process* (New Haven, 1921), pp. 104-105.
52. The Vienna Convention of the Law of Treaties, Article 53.
53. Statement of President Barack Obama while accepting the Nobel Prize at Oslo, *The Times of India* (New Delhi), December 11, 2009.
54. Michael Sheehan, *The International Politics of Space* (London: Routledge, Taylor and Francis Group, 2007), p. 56.
55. Ibid., p. 27.
56. Peter L. Hays, *Space and Security-A Reference Handbook,* ABC-CLIO, LLC (Santa Barbara, 2011), pp. 7ff.
57. This project was taken up under the aegis of the "Agreement Concerning Cooperation in the Exploration and Use of Outer Space for Peaceful Purposes" signed between the US and USSR in May 1972. Many observers then were less sanguine and less optimistic about such cooperative efforts.
58. Sheehan, n. 54, p. 65.
59. The Constitution of the USSR, 1977, Article 28.
60. Pravda, April 13, 196,1 cited in Vereshchetin, n. 9, p. 211.
61. Treaty on Principles Governing the Activities of States in the Exploration and Use of Outer Space, including the Moon and other Celestial Bodies, 1967. It entered into force on October 10, 1967.
62. Sheehan, n. 54, p. 183.
63. Pervasive cooperation achieved so far certainly appears beyond coincidence and is advertant and volitional. Details of actual incidents of cooperation have been discussed earlier in the chapter.

6

Mining of Asteroids: A Legal Analysis for Effective Governance

Introduction

Humanity has industrialised fast to deplete its natural resources with indiscriminate usage, thus, losing on sustainability. This dearth has led humans to desperately look for resources anywhere and everywhere. The economic necessity for new source deposits is obvious and needs no vindication. Alternatives have been sighted on the celestial bodies in outer space that are so rich in minerals and metals Thus, the entire universe becomes our provenance of natural resources and this reservoir which appears nearly limitless and inexhaustible has been drawn into the domain of utilisation and consumption by humankind. This fact had been surmised for long but has lately been confirmed by spectrometry studies and remote sensing data analysis.

Further, appropriate technology has been developed only recently and its applications made feasible. The viability of mining celestial bodies has now been proved from many aspects, including docking procedures, technical extraction capability, facilitation of transportation, and economic profitability. Therefore, near-exhaustion of mineral resources on the earth and their infinite availability as ore deposits on almost all the celestial bodies, particularly within the solar system, is soon expected to give birth to a new industry of space mining to tap mineral treasures from extra-terrestrial natural reserves. The ores from outer space are assessed to be rich in common utility minerals having a high element of currently precious metals like silver, gold and platinum.

It is, therefore, natural for the corporate sages of the business community to show keen interest in this space activity. A commercial conglomerate has harnessed billions of dollars to invest in space mining ventures. This corporation, styled Planetary Resources Incorporated, has obtained state-of-the-art technology and trained manpower to exploit, to begin with, an asteroid close in orbit to the earth. The overall risks as calculated are surmountable and within safety margins; and the economics of the venture assures rich dividends. The mining corporation has, therefore, made an announcement of its intention to privately prospect and exploit the natural resources available on the surface of, and inside, the asteroid Eros for commercial purposes. This is the harbinger of what is to follow. The race to competition does not seem far behind.

The seriousness of this intention leads to the question of the legal permissibility of commercial mining activities on the selected asteroid and other celestial bodies. This assumes importance because the legal regime of outer space is at variance with that of the planet earth in many substantive aspects. The activities and relationships in outer space are broadly governed under the international law but superimposed with the Outer Space Treaty[1] that declares that outer space "shall be the province of all mankind" and national appropriation in any manner or by any means is prohibited in outer space. However, space activities by non-governmental entities are permitted, *albeit* under national supervision and responsibility.[2]

Another legal instrument that can throw light on this issue is the Moon Agreement.[3] This document in its Preamble recognises that "benefits… may be derived from the exploitation of the natural resources of the Moon and other celestial bodies." Scientific investigation into outer space has proved the truth of this statement by mapping the rich reservoirs of mineral wealth and actual mining now is a hard reality. On the other hand, global mineral resources have been depleted to an alarmingly low level and future scarcity stares us in the face. New sources, due to dire necessity, needed to be searched for anywhere and everywhere. The moon and celestial bodies offer an abundant availability and benign opportunity. Man, as yet, believes that he is supreme and the entire universe is at his command and solely for his benefit. It is, therefore, natural that humankind is at the threshold of resorting to exploitation of the material resources of outer space.

The Moon Treaty also reiterates that "the moon shall be the province of mankind" and allows "activities in the exploration and use of the moon anywhere on or below its surface."[4] This permission can be logically extended to other celestial bodies, including asteroids. Hence, the benefits of such activities must accrue to humanity i.e. "all countries" in an equitable measure under a regime of common heritage of mankind to satisfy the avowed objective of benefit to humankind.

The existing legal provisions are inadequate and inhibitive for the specificity of purpose. A fitting and pertinent legal framework, as envisaged in the Moon Agreement, must now be put in place to release this treasure trove for optimum use. Scientific progress has made it incumbent that legally binding amplifications are consensually evolved and promulgated before competition and chaos seeps in or lest it be too late to introduce a proper economic order of sustainable usage and regulated extraction with equitable sharing of the resultant benefits as per stipulated modalities. The urgency is stark and the task laborious.

Nature of Asteroids

Asteroids are in the news and are getting too close to the earth, thus, causing an anguishing scare for humanity as well as flashing their treasures to charm the business tycoons. Asteroids are small bodies called minor planets or planetoids that orbit around the sun, especially the inner solar system. These are smaller than known planets but bigger than meteoroids, primarily located in the 'main-belt' beyond the orbit of Mars. But some become "near-earth" objects that swing much closer to our planet, sometimes even within 100,000 miles or nearer. Asteroids "that have been nudged by the gravitational attraction of nearby planets into orbits that allow them to enter the earth's neighbourhood" are called Near-Earth-Objects (NEO).[5] There are three groups of near-earth asteroids.

These stray passes have elicited great interest from scientists in different specialisations to study their orbital perigees, gravity parameters and material composition. As a result, some space organisations have landed robotic probes on asteroids, like the US landing on Eros for exploration and investigation. NASA is also planning to put a team of astronauts on an asteroid up to three million miles away from the earth by end 2020.[6] Japan is also reportedly planning to launch a humanned mission to "rendezvous with an asteroid up to three million miles from the earth." It is, of course, a risky affair due to the non-existent gravity, yet the astronauts may stay for a month on the asteroid. For this mission, a specialised NASA team is being trained to explore taxonomy and definitive mineralogy of the asteroid. The round trip may take about a year.

The Population of Asteroids

It has been believed for a long time that asteroids are low density rocks with almost non-existent gravity. Perchance, these broke away from the mother landmass and happen to be orbiting around the sun but at times the orbit comes perilously close to the earth to cause alarm. In a recent instance, an asteroid space rock (designated 2002 AM 31) zoomed past the earth on July 22, 2012, orbiting at 13.7 moon distances. It qualifies as a Near-Earth Asteroid (NEA) and is estimated to have been 2,000-4,500 feet wide, equivalent to the size of a city block.[7] In a similar incident, asteroid LZ1, NEO, unexpectedly dashed past the earth in mid-June 2012. Such occurrences are not rare, though they are of irregular frequency.

The known population of asteroids is numerous and still more are being discovered.[8] According to reliable estimates by astronomers, approximately 9,000 are orbiting in the solar system, generally in clusters called belts that have been classified into families. Of these, 1,500 are considered within relatively easy reach. Some of these have been individually identified and named or designated by numbers or both. Vesta is, as yet, considered to be the second biggest rock formation with origins similar to that of the earth. It materialised from condensation of diffused matter and its mass congealed from the same starting molten material as the earth. Its exploration and investigation can, thus, reveal many truths about the origin and formation of the earth and early universe.

At present, a NASA probe on planetary sciences, called DAWN, is orbiting around Vesta to study its spatial characteristics, physical composition and gravity parameters, among other tasks. The probe team is also mapping its resource deposits and has noticed crystalline structures of HED (Howardite-Eucrite-Diogenite).[9] It is surmised that Vesta has in its history passed through a violent collisional environment and its topography is pock-marked with the impact of the bombardment of meteorite rocks. Its surface also shows an undulating topography with lines like that of basins.

Ceres, however, is believed to be the biggest asteroid in size within the solar system and as yet remains an unexplored mystery. It has not so far been approached due to its distance from the earth and other attendant hazards. There are also other asteroids and dead comets that are smaller in size like Rosetta, Mathilde, Eros, Gaspra, Steins, Ida, Ivor, Juno, Pallas, Aten, Amor, Alinda, Amalthea, Apollo and Itokawa, to name just a few. Asteroid Arjuna, however, is more earth-like. It orbits in low inclination of low eccentricity and it has an orbital period of the earth's one year. A major concentration within the solar system is believed to be around the planet Jupiter.

There are many belts of such celestial bodies which are called families and are classified as such. The major NEA families are the Trojan asteroids, Cybele family, Hilda family and Hungaria family. The specific asteroids of interest to us are those orbiting in near-earth perigees or that are rich in mineral deposits of precious or scarce metals. Asteroid Psyche has been discovered to be a treasure trove of iron and nickel. Another asteroid, Nereus, has very low gravity, almost zero, but is believed to have an abundance of precious ores.

The Composition of Asteroids

Knowledge gained from robotic visitors so far is promising for the exploitation of celestial natural resources rich in rare metals and minerals. Technology and harvest prospecting of the quality and quantity of ores have prodded business houses to consider their economic viability and take calculated risks in excavation. Planetary Resources Incorporated is one such corporate entity that has mustered enough financial and technical resources to initiate the process and made known its intention of mining of asteroids in the near future. Another budding corporation, with similar goals and objectives, is Asteroid Enterprises Planetary Ltd. The proposal appears promising from many angles though it needs billions of dollars in investment and a multi-decadal gestation time to fructify.

That these bodies are rich in many raw materials has been well known to geologists and physicists for a long time, though recent probes have established the specifics of minerals and volatiles. Minerals that can be mined or extracted include iron, nickel, and titanium for construction. Precious metals like platinum, gold and cobalt can also be extracted. Apart from these, water can be mined from ice and can be economically broken down to liquid oxygen and liquid hydrogen that can be accumulated for storage. While oxygen can help to sustain the lives of the prospector-astronauts on the site, hydrogen can be used as rocket fuel for *in-situ* utilisation, to refuel on-journey vehicles, stocked as local tankage or as orbiting propellant depots. The ideas, though ambitious on economics, yet appear viable.

Incidentally, it would appear of interest to mention that planets have been spotted, observed and discovered which are super-rich earths made of diamonds that are still in the process of morphosis. In this connection, it seems equally pertinent to inform that recently a team of Indo-French astronomers[10] have discovered a planet twice the size of our own earth, orbiting a star visible to the naked eye that is largely composed of diamond.[11] The rocky planet, called '55 Cancri e', orbits a sun-like star in the constellation of Cancer and is moving so fast that a year there lasts a mere 18 earth-hours. Its radius is twice that of the earth and its mass is eight times greater. It is also incredibly hot, with temperatures on its surface reaching 3,900 degrees Fahrenheit (2,148 Celsius). The surface of this planet has a fundamentally different chemistry and is likely to be carbon-rich and covered in graphite and diamond. It is estimated that at least a third of the planet's mass, the equivalent of about three earth masses, could be diamond. The planets, in fact, are much more complex entities and future explorations may strive to prospect and discover such enviable bodies within our own galaxy.

In fact, all the gold, cobalt, iron, manganese, molybdenum, nickel, osmium, palladium, platinum, rhenium, rhodium, ruthenium, antimony, zinc, indium and tungsten that are mined from the earth's crust, and are essential for economic and technological progress, came originally from the rain of asteroids that hit the earth after its crust cooled and stayed embedded near the surface.[12] This is because the earth's massive gravity pulled all such native heavy siderophilic (iron-loving) elements into the planet's core during its molten youth more than four billion years ago. This left the crust depleted of such valuable elements and whatever remained has been indiscreetly mined. With globally asymmetrical industrial usage, such assets have been nearly exhausted while the needs of developing nations are escalating. Hence, the compulsion to look out for extra-terrestrial sources.

The search for alternative provenance has been made practical by Planetary Resources Inc with the proposed project of mining of asteroids and to begin with, of Eros. Hence, actual mining, in whatever manner, of celestial bodies appears just round the corner. It is sure to become a reality sooner than expected. The legality of such activities and that too by a private enterprise for commercial purposes remains moot and highly suspect. Only a contrived legal casuistry can justify such moves or uphold such motives. It is, therefore, proposed to discuss, in this chapter, the legal permissibility of such commercial activities on the celestial bodies in outer space.

Economics of Asteroid Mining

The economics of mining of asteroids is not simple and requires factoring in of a lot of parameters, whether for its spatial characteristics to ensure the safety of miners, the velocity of the celestial body, the distance of the target from the earth, commercial utilisation of resources *in situ* and expenses on transportation of raw or processed material to the earth or any other stocking depot. State-of-the-art transportation technology is not very promising yet and needs to work on simpler logistics and cost reduction.

Such ventures may also have a significant impact on the environmental conservation of the earth and sustainability of planetary resources. For these very reasons, the mining of Apollo and Amor asteroids has been under consideration since 1977 though it was expected to provide human civilisation with the potential to overcome the resource crunch and supply additional energy sources, thus, earnestly hoping to add millions to the global gross product. Regrettably, no tangible results have materialised so far.[13]

Further, some scientists and techno-economic analysts view the celestial mining plans with scepticism and do not find these cost-effective in a future scenario, with a paranoia about lowered gold and platinum prices consequent to a glut in the market. Thus, every aspect deserves the fullest consideration whether for radical reduction in space transportation costs, physical sustainability of equipment and installations in gravity-free conditions, sustainability of resources in a fragile environment or the ultimate motive of profitability from the total project.[14]

Asteroid Selection

Selection of the appropriate asteroid for mining operations depends on several techno-commercial factors. The most important factor to consider in target selection is orbital geometry and allied characteristics. These include, in particular, the change in velocity for hazard-free physical activity while higher velocities can facilitate planetary-assisted faster trajectories for cost reduction. The second consideration is travel time to and from the target which constitutes a major component of retrieval costs. Thus, these parameters make asteroids in low-earth orbit attractive propositions for profitability. These are the considerations of asteroid logistics.

Further, the ensuing developments of asteroid-orbit manipulation infrastructure could offer an irresistible return on investment by altering the trajectory of bodies to the nearby interplanetary space in low-earth orbit or achieve proximity to the moon. Theoretically, some asteroids can be inter-spatially taken down on the moon with the help of nukes and SOPL (Solar Orbital Power Lines) which can be fired to usefully adjust the orbit of the asteroid.[15] But such antics are endemic of collision risks and, in some eventualities, the results could be catastrophic. Astrophysicists Carl Sagan and Steven Ostro have raised valid concerns about artificially altering trajectories and have suggested instituting stringent controls on the misuse of orbit-engineering technology.[16] The point made deserves the fullest consideration and urgent attention.

The third element in asteroid selection is its native material composition and geometrical shape. Asteroids are generally clumpy and irregular in form, with geo-spatial asymmetry. Therefore, relatively small metallic asteroids of one mile diameter are the preferred population. For example, the asteroid '16 Psyche'[17] is a siderite and is believed to contain 1.7×10^{19} kg of nickel–iron, which could supply the world production requirement for several million years. A small portion of its native material could even turn out be precious metals. In another study, Planetary Resources Inc has assessed for eventual mining a 30-metre-long asteroid which could yield gold and platinum worth US$ 25-50 billion.[18] Indeed, a colossal net worth of the project that would be rather difficult to resist.

Other factors governing selection are the extent of the extracted materials' local disposability or *in situ* processing or utilisation to reduce the payload of shipment to the earth. It is desirable that most of the extracted material must be expended locally or processed to reduce its bulk. Therefore, near-earth asteroids are considered potential candidates for mining missions because their nearby location makes them greatly attractive, reducing the economic cost of the propulsion energy requirements into the earth's orbit. A suitable target on these parameters could be 4660 Nereus that has very low gravity compared to the moon.

Mining Considerations

Mining operations would require robotised excavatory equipment with special installations to handle the extraction; and furnaces and machinery for processing of mineral ore in outer space. The machinery will need to be anchored to the body of the asteroid provided it is rigid enough, but once installed, the ore can be moved about more effortlessly due to lack of gravity. Docking with an asteroid can be done using a harpoon-like process, where a projectile penetrates the surface to serve as an anchor, and then, an attached cable can winch the vehicle onto the surface.[19] The procedure, while being rather risky and hazardous, is technically achievable with due regard to the safety of the astronauts and miners.

Once the process of mining starts, there are two options for the disposal of the extracted ore. The first option is to bring the mined asteroidal material in raw form to the earth for refinement and commercial disposal. This is an expensive proposition for haulage to the earth. The second option is on-site processing so that the bulky waste is trashed on the rock body itself and the small quantity of precious extracted materials is transported to the earth. The extraction procedures may also release propellants, allied energy sources and other byproducts.

On cost-benefit analysis, this option may save on recurring transportation costs but processing machinery and ancillary facilities, as capital investment, would need to be installed on the asteroid which may involve heavy outlays. The installations may be automated and robot-friendly or manually operated with human presence. Incidentally, automated systems would be much less expensive to build and deploy but manual systems are easy to operate and provide for trouble-shooting. The comparative cost-effectiveness of alternatives will, of course, have to be evaluated in detail for a preferred solution.

Material Harvesting
Material extraction, also called harvesting, is the substantive part of the project and can be achieved by different methods. The first is strip mining in which the material is scraped off the surface of the celestial body and repeated in successive runs till it is exhausted or the ore quality is diminished. There is strong evidence that many asteroids consist of rubble piles that make this approach possible. The second method is shaft mining. In this method, a mine can be dug into the asteroid and material extracted through the shaft. This, however, requires precise knowledge to engineer accuracy of astro-location under the surface regolith and a transportation system to carry the desired ore to the processing facility or to the earth centre.

The third method is specific to excavation of ores with high metallic content that can be ground to loose grains that can be gathered by means of a magnet and transported to the processing facility. Another material-specific method is heating. This is useful in collection of volatile materials like water and gases. Here heat can be used to melt and vapourise the matrix.[20] The last method is to establish a complex automated factory to produce self-replicating machines. The power of self-replication is compelling and it would consume large quantities of harvested asteroidal material. This has been facilitated by technological progress in miniaturisation, nanotechnology, materials revolution and additive manufacturing. NASA has mooted this concept in seminal form but studies assure of its feasibility.[21] Whatever be the method of harvesting, asteroids offer treasure beyond count, power beyond measure and resources beyond imagination. Humanity is only now awakening to the riches of space.

Claims to Celestial Bodies

From time immemorial, man has vainly desired to personally possess the heavenly bodies and the moon has been the symbol of the ultimate wish. Eastern civilisations have personified this supreme desire; for example, Indian folklore is replete with stories of '*Chanda Mama*', referring to the moon as maternal uncle. There is a tale from Hindu mythology that Lord Krishna, as child stubbornly insisted on possessing the moon. The parents put some water in a flat vessel and showed him a reflection of the moon to pacify him. The moon, nevertheless, remained a mystery with mythical legends but today we know a lot about our favourite natural satellite.

Another personification of this unfulfilled wish is evident in the tradition of naming children on celestial stars and planets. Common names are Ravi, Suraj, Bhan, Aftab, Khursheed, and so on, which all signify the sun. Similarly, names like Chan, Chand, Chandrika, Sasi, etc take after the moon. There are other popular names taken from other planets like Mangal, Budh, Shani, etc. This custom cuts across almost all communities and tribes in India, irrespective of religion.

Claims to celestial bodies on frivolous grounds or as half-serious jokes have been attempted for millennia yet none has stood the test of law with proven ingredients and substantiated evidence or the ravages of time to be recognised in custom by communities. On the other hand, ownership and possession is an innate desire of humans and a part of their instinctive nature. In fact, it is the manifestation of this instinct that has impelled mankind to growth and prosperity. Thus, private ownership or *jus in re,* meaning the right to property provides the correct stimulus for material progress. Socialism, on the contrary, induces sluggish material production and impedes economic development on both personal and community levels.

Mere claim to extra-terrestrial bodies, believed to exist in the cosmos, is not enough proof for ownership. This establishes only intention or animus and is not sufficient for the purpose. For ownership of property, the corpus must be acquired in the first instance and, at the same time, its actual possession with virtual control over its usufructs as well as absolute right to disposal of the same must vest with the owner. *Jus ad rem* without possession would be inchoate and incomplete right *ipso jure*. Of course, on these parameters, a first landing on a celestial body like an asteroid, with appropriate intention, even for a probe or exploration, may fulfill substantive conditions of an ownership claim on extra-terrestrial land and yield a right to mineral resources. This could be valid in international law of territory acquisition but not under the edicts of space law.[22]

The moon has been sold many times over and buyers are confident of obtaining possession one day, however remote and distant. In the recent past, there was a seller of plots on the moon advertising vigorously in the US and to other gullible buyers. Some Indians have also bought these.[23] *Caveat emptor*: the seller neither has the legal right to the property on offer nor has yet taken possession of the same. Without both, the offer does not stand on contract law or make for legal transfer on laws of property. Sheer payment of consideration for an uncertain and speculative promise in the future does not give it the validity or nobility of the contract.

The celestial realty also cannot be transferred by a gift deed or acquired by lease deed from private individuals. Such dubious transactions have happened unchecked only under the loose laws of a few states in the US because the OST, at best, is confusing and at worst, internally inconsistent on space property rights. But *qui tacet negat*, which means the silence of authorities, cannot be treated as acquiescence to the fact nor can an adverse inference be drawn. Therefore, a heated debate on the ownership of space property and realty rights is likely to take place in the not too distant future.[24]

Over the last century, there have been many such cases of claims on extra-terrestrial bodies, some serious and determined, others sheer fiction or a legal joke but almost all are based on the trivial argument of unclaimed property, and deserve to be debunked as fictional episodes, void in law. These instances not only highlight a dire need for regulatory laws that are explicit and unequivocal but also necessitate a mechanism to curb and punish such frauds on gullible people. Proper propagation of rules in the public interest may ameliorate the situation considerably.

Let's illustrate with a few examples. First, in 2002, Virgiliu Pop, a researcher at the Romanian Space Agency, proclaimed ownership of the sun with an online register and threatened owners of the moon and other extra-terrestrial property owners like Bill Hope and Gregory Nemitz with legal notices for charges relating to the use of sunlight. He has later accepted in his book that a claim is just a claim unless supported by an ownership deed and its registration does nothing more than certify the fact of its having been staked. Nevertheless, it was intended as a joke and to publicise the hollowness of similar claimants.[25]

Real estate on the moon has been on sale in the US from as early as 1890. Lunar plots, defined by astral longitude and latitude grid references, and other attractions have been peddled by unscrupulous sellers to the gullible and credulous buyers because it was easy to get a charter for the purpose in the state of New Jersey. On June 15, 1936, Dean Lindsay, a Georgian, averred before a notary ownership of the 'Lindsay Archipelago' comprising islands of space denoting celestial bodies. He anticipated commercial value for these rocky formations and laid his right over them as unclaimed bodies, but, regrettably, he did not visualise the claim to the void of outer space.[26]

In 1948, James Thomas Mangan claimed the 'Island of Space' to establish the Nation of Celestial Space, in short, "Celestia" on the plea that humankind on earth travels 1,500,000 miles each day through the territory of Celestia. On December 29, 1948, Celestia formally applied for membership of the United Nations. It was rejected under Article 4 (1) of the Charter of the UN alleging inability to discharge incumbent obligations. In 1956, Mangan became critical of the UN and blamed it for being "legally impolitic and scientifically impolite." Later, he declared that Celestia was the only competent entity to frame laws for outer space and ordained all nations on earth to obey and abide by this declaration. He also proclaimed that the entry and presence of artificial man-made satellites in outer space tantamount to trespass violative of its state sovereignty.[27]

In 1952, the Berkley Science Fiction Fan Club filed a legal claim for a triangular lunar area covering three craters — Ritter, Manners and Sabine — and sent a notice to the UN to issue the patent and title. This claim was based on discovery of sylvanite in the region and was publicised in the media.[28]

In September 1954, Planet Mars Development Corporation was formed by three businessmen in Arkansas who offered to convey the title to competent persons for suitable remuneration. Following the same *modus operandi*, in November 1955, Robert Coles constituted the Interplanetary Development Corporation to offer celestial plots at one dollar an acre against a "General Quit Claim Deed."[29] Another important assertion was made in 2005 by Hope who operated "the Lunar Embassy" and boasted of having solicited 3.6 million extra-terrestrial property clients in 181 countries during three decades of his 'unreal' estate business.[30] Thus, fraudulent sale of "unreal estate" has been a near-monopoly of the US.

Outside of the US, in 1953, Janeiro Gajardo Vera, a Chilean lawyer, claimed the moon. He approached local property registration authorities and was directed to make a public proclamation of his right to the subject property. As per requirement of the applicable laws, he published his claim three times seeking adverse claimants. Fortunately, for him, no objections were received by the authorities and the moon was registered as his property and the mutation endorsed in his name. The true motive of Vera, however, was to obtain membership of a club which had been refused for he owned no real estate.[31] One can only speculate if this ruse helped him achieve his purpose.

There is, however, a specific case of a property claim to an asteroid called Eros. The US space agency NASA had made an automated robotic landing of the NEAR-Shoemaker probe on Eros in 2001. Gregory Nemitz, a space activist, operating an online database styled Archimedes Institute, who claimed ownership to this property, filed a bill on NASA and the State Department for parking fees relating to the landing of the US satellite on Eros. It was to be expected that the US State Department and NASA would repudiate the claim and reject payment of the bill for parking charges.

Nemitz then filed a law suit for his claim on parking fees in the US 9th Circuit Court of Appeals. The plea taken was that the OST prohibits national appropriation but not constructive individual rights thereto. The contentious litigation was entertained but on arguments, the suit was dismissed because the petitioner was not able to prove his case. He could neither produce legal documents of ownership rights purchased from the seller nor substantiate taking over of actual and physical possession for the property named Eros nor the exercising of his effective control thereon. Nemitz failed on both, germane law as well as possessory evidence.[32]

A Legal Analysis

The corpus of space law has evolved over just half a century and is still metamorphosing. The instruments are a composite of treaties, agreements, UN declarations and guidelines. It is partly *pacta* and partly soft law but it lays down general and broad mandates and lacks in details. It, thus, has grey areas of doubt and reveals chinks in interpretation. The legality of mining of celestial bodies, including asteroids, is one such subject dotted with holes and tears. Constructions placed on the Outer Space Treaty almost border the absurd though the logic appears robust and flawless. Sometimes, syllogistic fallacies do come to the surface yet confusion reigns and vested interests are likely to prevail. The situation is still under control and is redeemable. Ambiguities can be identified and cured while inconvenient specifics can be addressed and promulgated for compliance.

The legality and consequent permissibility of mining celestial bodies, including asteroids, can be derived from two major instruments, the Outer Space Treaty and the Moon Agreement. It is also pertinent to mention that the OST contains the fundamentals and is generally treated as *grundnorm* of space law. The OST has been so widely accepted that it has been exalted to customary space law that is universally binding, even on non-parties. The Moon Treaty is more specific to celestial bodies regarding their exploration, use and exploitation of available resources for the benefit of all mankind. But, on the other hand, it is a relatively weak instrument, with few votaries and rather limited adherence. The rest of the space law is not directly germane to the subject except for minor relevance or indirect reference. So the template is clear for analysis and verdict.

The Outer Space Treaty

The Outer Space Treaty is extraordinary in many respects and unique from the legal point of view. It enunciates the fundamental principles of space law that are novel if not really new. These are obligatory and binding not only on state-parties but all states internationally. Therefore the legal provisions of the treaty are mandatory for compliance whether in permissibility or prohibition of a space activity. This is the international law of outer space. There may be some grey areas on which the treaty is either not specific or is silent. These can be interpreted according to the canons of international jurisprudence, keeping the best interest of humankind in the right perspective. Indeed, the focus is clear and fixed.

The Preamble to the Treaty

Let's begin with the Preamble to the Treaty, which many may argue is not the legal text and, hence, merely mentions the statement of intent. This view may be technically correct but the Preamble truly embodies the best narration of human aspiration, that is, the welfare of all and the ushering in, and sustaining, of peace in the universe. The ideas are noble and the thoughts sublime. The humane aspect of the Preamble is, thus, truly appealing and overwhelming.

The Preamble, at the very outset, implicitly accepts that outer space, though still in major part a mystery, appears a vast reservoir of (mineral and metal) resources apart from diverse other benefits that can, in due course, accrue to humankind. Fortunately, it has become possible to explore, prospect and confirm their availability and abundance by remote sensing and on-site visits to celestial bodies and asteroids by robot rovers. And today, technology has been developed that has made it feasible to viably exploit these resources, at least in the near-earth orbit.

In view of this scenario, the Preamble begins by stating that the state-parties are "inspired by the great prospects opening up before mankind as a result of man's entry into outer space." Then it recognises "the common interest of all mankind in the progress of the exploration and use of outer space for peaceful purposes." And exhorts the states that "the exploration and use of outer space should be carried on for the benefit of all peoples, irrespective of the degree of their economic and scientific development."[33] The principles are laudable and reveal a slant towards a socialist culture, tempered with the ultimate humanitarian purpose. This viewpoint, though not new to law, has been, for the first time daringly expressed by the international community in unity and with unanimity.[34]

The above clauses are pregnant with dire uncertainties. On the face of it, the treaty accepts space activities by the states. In fact, at that time it was near impossible to imagine that private companies could muster such vast capital funds to invest in space activities and compete with the budgets of states. Hence, these provisions preclude private enterprise for individual profit and personal benefit unless the dividends are shared equitably among all mankind under some acceptable modality. But this would be hard to achieve because the capitalistic economic order is not built on altruism and charity. The incompatibility of espoused causes with world reality becomes evident. Thus, the need is for a change of mindset.

Province of All Mankind

The Outer Space Treaty enshrines the principle that outer space, including the moon and other celestial bodies are "the province of all mankind." This concept is similar, if not congruous, to the doctrine of common heritage of mankind from the Law of the Sea. Though similarities become apparent, there are subtle differences of significance. For example, heritage has an implication of legacy as *res communis omnium* and carries an inherent possibility of divisibility and apportionment and this option, whatever be its legality or illegality, has been exercised by the US in the extraction of manganese nodules from the seabed.

The text of the treaty reads, "The exploration and use of outer space, including the moon and other celestial bodies, shall be carried out for the benefit, and in the interest, of all countries, irrespective of their degree of economic or scientific development, and shall be the province of all mankind." It becomes explicit that outer space belongs to mankind as a whole as *res publica* laced with public trust doctrine or *res communis humanitatis* with endemic indivisibility. A comparable Indian concept of such property rights would be *shamlat* or *panchayati* land which is open to the entire village as free common grazing ground. Here the connotation of mankind includes future generations also and their inherent heir-apparent right to 'the province' of opportunities and resources.

Further, the treaty "recognises the common interest of all mankind" in the Preamble and directs that space activities "...shall be carried out for the benefit, and in the interest, of all countries..."[35] These provisions clearly point towards the stakeholding of all countries in the beneficial accruals of any variety from outer space. Therefore, shares of profit consequent to commercial activities in outer space and on the celestial bodies must legitimately flow to all countries in reasonable proportion. The population of beneficiaries, thus, does not permit national or private ownership, excludes private profiteering and imposes a social responsibility to genuinely divide the economic benefits in equitable proportion. The modalities and scope of sharing may yet have to be devised amicably among the states, nevertheless, the mandate of sharing the benefits is obligatory and binding; it is not merely notional or promissory.

Non-Appropriation in Outer Space

As a corollary to the doctrine of outer space as the province of mankind there is another provision in the treaty that supports and reinforces it in tandem. This is the principle of non-appropriation in outer space. The text of the treaty states, "Outer space, including the Moon and other celestial bodies, is not subject to national appropriation by claims of sovereignty, by means of use or occupation, or by any other means."[36] This means that there shall be no national borders or flags of sovereignty on the moon and other celestial bodies. Ideally, it secures outer space, including the celestial bodies, as a unified expanse of '[Hu]Mankind Commons.'

The concept of non-appropriation is amplified by another clause of the treaty, "Outer space, including the Moon and other celestial bodies, shall be free for exploration and use by all States without discrimination of any kind, on a basis of equality and in accordance with international law, and there shall be free access to all areas of celestial bodies."[37] The assured freedom of exploration and use and free access to all areas of the celestial bodies guarantees that there would no boundaries, hindrance or let to this right of activity and necessary movement in outer space. It promises an egalitarian regime of operations in outer space with no rights to appropriate any part or portion of outer space or any celestial body for proprietary or exclusive use.

The abovementioned right is, however, regulated in that states "shall carry on activities in the exploration and use of outer space, including the Moon and other celestial bodies, in accordance with international law, including the Charter of the United Nations..."[38] The applicability of international law, including the Charter of the United Nations, complicates the scenario and solicits an interpretation. It will be seen that traditional international law on some points comes in conflict with the tenets of space law. For example, territorial sovereignty and acquisition of territories by different methods are time honoured concepts of international law but space law outright negates these. In a way, however, space law acknowledges collective sovereignty and universal community ownership of all mankind, with a substratum of common interest, over outer space and the celestial bodies enjoining responsibility towards sustainability. Nevertheless, jurisprudence dictates that *lex specialibus* supersedes *lex generalis* and, hence, to the extent of inconsistency, space law shall override and duly operate.

The basic space law ordains that states cannot carry territorial sovereignty to outer space and the celestial bodies. Further, a general inference can be logically drawn that there are no private property rights in outer space and on the celestial bodies. The treaty considers and recognises only states as participants and their activities in the exploration and use of outer space as legitimate and licit. There is, thus, no scope for any wider or liberal interpretation under this provision at this juncture under the existing legal scenario.

Freedom of Access and Investigation

Another cardinal principle of space law contained in the Outer Space Treaty stresses freedom to all states for scientific investigation and access to all parts of outer space, including the celestial bodies. The treaty provision states, "Outer space, including the Moon and other celestial bodies, shall be free for exploration and use by all states without discrimination of any kind, on a basis of equality and in accordance with international law, and there shall be free access to all areas of celestial bodies."[39] The treaty ensures this "freedom to all States" and at this stage, makes no authorisation to delegate such freedom to private enterprise or individual entrepreneurs.

There is a similar "...freedom of scientific investigation in outer space, including the Moon and other celestial bodies, and States shall facilitate and encourage international cooperation in such investigation."[40] This satisfies the quest of humans for novelty and curiosity. The freedom is assured only to states and for scientific investigation and it gives no licence for reckless investigative activities that may cause wanton damage or harmful destruction. The need for sustainability of the pristine environment is clearly implicit in this treaty clause.

This freedom also carries a correlative duty to respect the corresponding right of other states and not to disturb, or interfere in, their legitimate operations. This appears a reasonable restriction to the absoluteness of the freedom espoused in the treaty. The text guides that "...States...shall conduct all their activities in outer space, including the Moon and other celestial bodies, with due regard to the corresponding interests of all other States-Parties to the Treaty."[41] Equality of states, mutuality of interests and reciprocity of responsibility secures freedom for the states under the treaty.

The connotation of state does not include private commercial organisations or individual entrepreneurs who would normally operate under business ethics and be motivated by profit concerns. Hence, the legitimacy of mining activities on asteroids by planetary resources is in doubt. Further, by such relaxations, relationships may get skewed because the latter may not show the same regard or sensitivity that may be expected of, or be voluntarily forthcoming from, the states as per international norms of state practice.

Hints of Permissibility

Use for Peaceful Purposes
The Outer Space Treaty permits in its Preamble the "...use of outer space for peaceful purposes." And in another auxiliary clause in Article IV of the treaty, it enjoins, "The Moon and other celestial bodies shall be used by all States-Parties to the Treaty exclusively for peaceful purposes."[42] This combination of clauses highlights that the nature of activities in outer space

is to be for peaceful purposes. It, thus, raises two issues: first, that states are permitted 'the use' of outer space and celestial bodies; and second that it should be exclusively for peaceful purposes.

First, use of outer space and the celestial bodies is permitted to the state-parties, but their 'use' can take various forms, methods and techniques. For example, the use could be visual, observatory and non-invasive like remote sensing or broadcasting. Or it can be possessory and invasive yet non-destructive like satellite landing and parking, picking of soil samples, or tourist habitations; or it could be intrusive, excavatory and destructive with harmful consequences like mining and other experimentation.

The second aspect relates to the nature and scope of peaceful uses. The treaty only prohibits "nuclear weapons or any other kind of weapons of mass destruction" from placement "...in orbit around the earth...install such weapons on celestial bodies or station such weapons in outer space in any other manner."[43] And states "undertake not to place" any such objects in prohibited places. So far, it is loud and clear, but the treaty nowhere defines peaceful activities by character or traits. One interpretation could be that what is not prohibited is permitted, hence, can be deemed peaceful activity; but the validity of such an assumption is illusory.

Be that as it may, it appears certain, beyond reasonable doubt, that the mining of asteroids is a peaceful activity of a high order. Therefore, if government agencies undertake this activity in outer space, it would attract no illegality or ban violation of any kind. Further, the treaty permits that "[t]he use of any equipment or facility necessary for peaceful exploration of the Moon and other celestial bodies shall also not be prohibited."[44] However, a dilemma remains whether the miner-state can claim property ownership rights to the excavated material with prerogative to its commercial disposal *in situ* or after transportation to its jurisdiction on the earth.

Public-Private Cooperation

But scenarios change fast. Today, technology has progressed tremendously to realise the visions of yesteryears. The reach of science and advancement of technology have stretched beyond human imagination. The moon is no distant destination, while Mars and Jupiter appear next-door neighbours.

Deep and deeper space probes are being planned and astronauts trained for long duration stays in outer space. These efforts are now being supplemented and gradually supplanted by non-governmental entities. The shift from governmental monopoly to sharing of missions with the private sector is clearly discernible and is a harbinger of public-private cooperation.

Alongside government players, private enterprise has also joined the projects and has grown into its own, financially and technologically. Some of the space activities that have been commercialised, like broadcasting and remote sensing and space tourism are already being progressively monopolised by private corporations and entrepreneurs. Private commercial operators are already here trying to broaden their reach and fervently hope to stay and prosper. It seems they now deserve some legal space for activities in peaceful uses. But the law fails them in their endeavours and has to be contrived to gain legitimacy and extract permissibility.

An indirect mention of space operations by non-governmental entities under the provision for international responsibility of states for national activities lends some solace. The text of the treaty reads, "States-Parties to the Treaty shall bear international responsibility for national activities in outer space, including the Moon and other celestial bodies, whether such activities are carried on by governmental agencies or by non-governmental entities…are carried out in conformity with the provisions set forth in the present treaty."[45]

This Article of the treaty further amplifies, though in a different context, that "[t]he activities of non-governmental entities in outer space , including the Moon and other celestial bodies, shall require authorization and continuing supervision by the appropriate State-Party to the Treaty."[46] This means that private enterprise cannot enter the space arena for operations without permission from the state and the government is obligated to monitor such activities and audit their safety for it bears the burden of *vinculum juris* (legal liability) towards international responsibility for any damage caused by such activities in outer space, in the air space, and on the earth.

Therefore, the treaty indirectly admits of the possibility of space activities being undertaken by the private economic sector or even individuals as legal entities, but under strict governmental care and control. In a way, it will be a progressive approach to reduce the scramble for natural resources on the earth and mitigate conflict for scarce materials by providing viable alternatives. In consequence, this viewpoint may facilitate and contribute to global peace.

This further prods us to an assumption that the consortium of planetary resources may have legal umbrage. Therefore, the corporation would certainly have obtained appropriate clearances and approvals from the US Administration under the domestic space legislation, namely, the National Aeronautics and Space Act, 1958, and other relevant municipal laws for their proposal of intent before initiating activities to ultimately mine the asteroid Eros. With the reelection of President Obama for the next term from 2013, NASA's plan for manned missions to the asteroid shall remain on the cards for execution as scheduled for 2025.

Not Prohibited is Permitted

Some jurists take comfort from an old dictum of law that what is not prohibited is deemed to be permitted as an adverse inference. This is a characteristic of criminal law codes which, in generality, list out and elaborate on prohibited actions and behaviour to be eschewed by the nationals and that in the event of violation or breach, it would attract the stipulated punishment. This method of enumerating aberrant behaviour applies to criminal law because the list of good conduct and acceptable acts would be too long to be comprehensive and, thus, make bulky reading. Hence, prohibitions are included in the code and what does not find mention therein is considered a lawful activity and, hence, not punishable. The logic is indeed simplistic and *lato sensu*; it does not find valid application in many other branches of law, more so international law.

This dichotomy may not always be exact and sacrosanct as, at times, it may be difficult to simplistically classify human behaviour on opposite sides. There would be grey areas of doubt because the contrary of false may not always be true, and *vice versa*. Hence, there is an inherent syllogistic fallacy in this form of bifurcatory logic. This line of thinking may not invariably lead us to the right answer which has to be specific and categorical.

Let's, therefore, look for what is permitted and patently legal under the treaty rather than take shelter under flimsy premises of tenuous validity. It is arguable that privated companies have a clear and legitimate right to operate in outer space and on the celestial bodies. The permission in the OST to non-governmental entities becomes vivid from the responsibility that the treaty attaches to the governments to regulate and supervise private activities by their juridical nationals. Further, there is not even a suggestion of prohibiting commercial or business activities in space.

An interesting example of such understanding of the law can be derived from the policy of the Government of the United States that has given a bold go-ahead to a US conglomerate, Planetary Resources Inc, for mining on an asteroid. This authorisation, it seems, has been given under the 10th Amenment which states, "...all powers not delegated to the Federal Government, nor prohibited by it to the States, are reserved to the States or to the people..." Thus, the right to mine in outer space, by direct inference, devolves on the people, though no claim to property title is tenable.

In the alternate, assuming if private enterprise and individuals are not properly authorised to undertake space activities or ownership of celestial bodies is specifically denied, the solution is not in placing devious construction on the law but to suitably amend the same to make it beneficent and transparent to be able to reap incumbent advantages so utterly necessary to continued human progress. The time appears ripe to shed prejudices, thaw frozen mindsets and strive for another wave of reforms in space law, preferably proactively, in consonance with the technological vision and to the satisfy growing needs of humanity. Wisdom lies in this option.

The Moon Agreement

Another international instrument of significance on the subject is the Moon Agreement. It is ,of course, more elaborate and detailed on specifics than the Outer Space Treaty yet it is also not comprehensive enough to provide direct and straight answers to all and sundry queries. Nevertheless, the agreement takes full advantage of the experience of the working of the Outer Space Treaty in its drafting and synergises the efficacy of the corpus of space law. This agreement admits of chinks in the law and the need for additional regulatory provisions as and when technology advances and man's reach to the celestial bodies becomes a safe routine and capabilities to exploit them assume diverse forms not envisioned so far. That stage has now been reached. Therefore, let's search for lawful permissibility of space activities by the private sector and enterprising individuals under the legal template of the Moon Agreement.

The Preamble to the Agreement

It is satisfying that the Preamble appreciates "...the achievements of States in the exploration and use of the Moon and other celestial bodies," and acknowledges that multifarious benefits "may be derived from the exploitation of the natural resources of the Moon and other celestial bodies."[47] The foreboding possibility and financial viability of such exploitation could cause a scramble for riches, leading to crass competition that may result in a chaotic economic order.

The negotiating nations were conscious as well as apprehensive of unregulated, unseemly developments in this sphere and, thus, desired "to prevent the Moon [and other celestial bodies] from becoming an area of international conflict."[48] To ameliorate the situation of scant and sketchy documents of space law, the states-parties realised "the need to define and develop the provisions of these international instruments in relation to the Moon and other celestial bodies, having regard to further progress in the exploration and use of outer space."[49]

The Preamble to the agreement reveals that it is a visionary document that envisaged myriad benefits from the natural resources of the moon and celestial bodies and that such development will be in the interest of mankind and for the welfare of humanity. Global resources were no longer sustainable and had been nearly exhausted. Therefore, an avaricious scramble for scarce minerals and metals, so vital for human progress, could be perilous.

This document recognised the alternative potential sources from outer space and attempted to supplement the existing corpus of space law by elaborating on the details and filling the specific gaps. The agreement, though it establishes a legal regime to regulate orderly exploitation of space resources, is still not comprehensive in the rules of governance for curbing violations. But it is, indeed, a wise step though a short one.

Applicability of the Moon Agreement

The Moon Agreement carries a special provision on its applicability. This clarificatory clause seems pertinent and specific. The text reads, "The provisions of this Agreement relating to the Moon shall also apply to other celestial bodies within the solar system...except insofar as specific legal norms enter into force with respect to any of these celestial bodies."[50] It is, thus, clear that it, rather realistically, applies to the celestial bodies within the solar system only and not, so ambitiously, to the entire universe. It is also pertinent to note that the Moon Agreement excludes applicability to the earth under the same Article and leaves governance of the earth through its own existing laws, customs and practices.

Unlike the Outer Space Treaty, it makes a sensible exception that its legal regime is not applicable to the earth.[51] This exclusion saves the agreement from a lot of confusion, and misinterpretation of its provisions. An allied clause further clarifies, "This Agreement does not apply to extraterrestrial materials which reach the surface of the Earth by natural means."[52] This implies that the agreement does not concern itself with natural activities in the solar system but only pertains to, and regulates, human exploratory activities and various uses of celestial bodies.

It has been discussed earlier that asteroids are rock formations orbiting around various planets of the sun and are, thus, admittedly, celestial bodies of our solar system. There is no denying that other solar systems would also be having similar orbiting bodies. Further, the above provisions of the agreement make it amply clear that by virtue of its applicability to celestial bodies within the solar system, it becomes directly relevant to asteroids and in regulating human activities on such bodies. Therefore, the Moon Agreement is the other fundamental international instrument from where to search for legal permissibility of mining on the asteroids.

Province of Mankind

The Moon Agreement reiterates this principle cherished in the Outer Space Treaty. In fact, here it is in a more elaborate and detailed form. This averment can be vindicated from the text of the agreement. "The exploration and use of the Moon [and other celestial bodies] shall be the province of all mankind and shall be carried out for the benefit, and in the interest, of all countries, irrespective of their economic or scientific development. Due regard shall be paid to the interests of present and future generations as well as to the need to promote higher standards of living and conditions of economic and social progress and development in accordance with the Charter of the United Nations."[53] The amplifications are explicit and laudable.

The concept of province of mankind outlined in the agreement puts emphasis on the use of moon resources for the collective welfare of humanity and to equitably improve the standards of living of mankind and all-round development as envisaged in the Charter of the United Nations. The best part is that the agreement does not restrict the benefits to *states-parties* only but pledges "in the interest of all countries." The altruism enshrined is commendable and should be emulated in other realms of activity as well. Another liberalisation of the concept relates to the concern for future generations, thereby, implying sustainability of resources. We only need to sincerely live up to these ideals.

Apropos mining of asteroids, it becomes amply clear that these are the province of mankind and cannot be individually or privately appropriated either as a rock body or for extracted material. It belongs to mankind collectively and has to be sustainable for future generations. And, lastly, the benefits must accrue to, and flow towards, all countries equally and equitably. Profits cannot be pocketed entirely by the private mining enterprise; if at all such activity is deemed permissible, it shall have to be proportionately shared globally, according to modalities to be negotiated.

Common Heritage of Mankind

The doctrine of Common Heritage of Mankind (CHM) is rather old and has been concurrently developed along, and gained acceptance, with the Law of the Sea. It has been adapted to cover the moon and celestial bodies in outer space. This concept had become relevant because of the abundant availability of metals and minerals deposited in the rock formations of celestial bodies and the progressive feasibility of their economic exploitation by technological advances. It also seems certain that governmental public funding may not be devoted to such ventures of commercial interest and private enterprise may enter this arena due to the lure of lucre. The corporate sector is, therefore, poised for this leap and is ready with the requisite wherewithal to make such exploitation a reality sooner than envisaged. However, one obstacle to this race is explicit legality under germane space law.

An analysis of the related provision in the agreement will clarify the situation. The agreement states, "The Moon [including other celestial bodies] and its natural resources are the common heritage of mankind..."[54] This clause means and reiterates the Outer Space Treaty, to aver specifically, "The Moon is not subject to national appropriation by any means of sovereignty, by means of use or occupation, or by any other means."[55] If we allude to the applicability clause under Article 1 of the agreement, the implication is that asteroids cannot be appropriated by any claim or by any means.

To clarify any ambiguity, the agreement further amplifies, "Neither the surface nor the subsurface of the Moon, nor any part thereof or natural resources in place, shall become the property of any state, international intergovernmental or non-governmental organization, national organization or non-governmental entity or any natural person." Any remaining doubts are dispelled by a subsequent clause that states, "The placement of personnel, space vehicles, equipment, facilities, stations and installations on or below the surface of the Moon, including structures connected with its surface or subsurface, shall not create a right of ownership over the surface or subsurface of the Moon or any areas thereof."[56] The common heritage of mankind is a concept of collective ownership which implies that everybody owns the property and, at the same time, nobody owns it, in the true sense, to make use of it individually or enjoy its usufructs. Thus, a Marxist-Socialist leaning is discernible in the concept. This is not encouraging because the fact that these have failed as economic systems is now a cliché. This may not be taken to imply that the alternative of capitalism is bound to succeed. This prospect is equally suspect for capitalism on earth has degenerated, and needs to be revisited, remedied and revitalised in consonance with contemporary work motivations and societal aspirations to make it suitable for the outer space milieu and exuding the hope of success.

In other words, it is *jus individuum,* an indivisible right of the public as a community. Its collective character is based on the model of trusteeship of property but it lacks the appointment of a trustee to take decisions in the best interest of the trust and its beneficiary, mankind. Thus, this *jus utendi freundi,* meaning that the right of liferent enjoyment of sustainable usufructs, lies dormant and held in abeyance. Perhaps, the United Nations is best suited for this task and can establish an appropriate organ or organisation equal to the assigned mandate of duty and responsibility. Perhaps, an idea of a World Space Authority can be mooted for this objective.

The agreement has envisioned a paradigm of CHM and has incorporated some guidelines within it for the utilisation of natural resources discovered on the moon and other celestial bodies, including asteroids. Under the agreement, "States...undertake to establish an international regime, including appropriate procedures, to govern the exploitation of the natural resources of the Moon... [when the same] is about to become feasible."[57] Man has pushed the frontier every day with success and the envisaged stage has been reached. The time is now ripe for framing of suitable regulations "compatible with the purposes" for sustained excavation and proper sharing of the proceeds over and above conscionable profits of production and reasonable rewards for cumulative efforts to making such exploitation practical.

The purpose and object of the proposed regime to be established shall include:

- The orderly and safe development of the natural resources of the Moon;
- The rational management of these resources;
- The expansion of opportunities in the use of these resources;
- An equitable sharing by all States-Parties in the benefits derived from those resources, whereby the interests and needs of the developing countries, as well as efforts of those countries which have contributed either directly or indirectly to the exploration of the Moon, shall be given special consideration.[58]

The proposed management regime highlights three aspects for attention. First, development and utilisation of natural resources shall be in a rational manner, keeping in view their sustainability. Second, profits from the venture shall be equitably and globally shared by all countries, irrespective of its being a space-faring country or not. Third, there shall be special consideration for the needs of the economically weaker nations as also for those who have bestowed funds and efforts in the cumulative exploration process as a result of which commercial exploitation became a profitable reality.

Notwithstanding the above advantages, the permissibility of exploitation by private enterprise remains a moot issue mired in controversy. Clarity is lacking and that is a dire disincentive to captains of business. The economics of industry is prodded by the profit motive and stimulated by an unencumbered legal environment for production. Both parameters are weak and suspect because profits are besotted with ominous risks and legality shadowed by a penumbra of controvertible doubts. Therefore, the legality of mining of asteroids by private industry remains plagued with uncertainty and apprehensions to become an enigmatic decision.

Freedom of Access

The Moon Agreement, in a similar vein as the Outer Space Treaty, assures freedom of access to all parts of the moon, even below its surface. In a way, this freedom is more expansive than that contained in the treaty as it permits digging to reach the subsurface in the process of exploration and use of the moon. Thus, the use can involve excavation and the agreement grants permission to move "freely over or below the surface of the Moon." This broader freedom should, by logic of applicability, rightfully extend to other celestial bodies and asteroids. A ray of hope is sighted.

The above proposition can be supported from the specific text of the agreement that reads, "States-Parties may pursue their activities in exploration and use of the Moon anywhere on or below its surface..."[59] These activities are further facilitated by authorising states to:

- Land their space objects on the Moon and launch them from the Moon.
- Place their personnel, space vehicles, equipment, facilities, stations and installations anywhere on or below the surface of the Moon...

and these...may move or be moved freely over or below the surface of the Moon.[60]

Just as every freedom is circumscribed by reasonable restrictions, so is this one. It also carries a set of curbs to ensure equal freedom to other parties operating in the field. First, "Activities of States...shall not interfere with the activities of other States... [and] where such interference may occur, the...concerned shall undertake consultations..."[61] This aspect assumes importance in respect of asteroids which are relatively much smaller bodies — some of these may not be able to accommodate two sets of facilities and installations by two competing mining firms. Consultations would be mandatory under such circumstances.

"To this end, all space vehicles, equipment, facilities, stations and installations on the Moon shall be open to other States-Parties...[on] reasonable advance notice of a projected visit...to assure safety and to avoid interference with normal operations in the facility to be visited."[62] Reciprocity and transparency are the obvious objectives. The agreement expects states to fulfill the incumbent obligation and provides for a procedure of consultation for the resolution of any controversy and for the ultimate dispute redressal.

The second restriction relates to ensuring that activities "in the exploration and use of the Moon are compatible" with the agreement and the latter is emphatic in maintaining, "The Moon shall be used by all States-Parties exclusively for peaceful purposes."[63] The directive is clear and explicit with no equivocation whatsoever. This implies that mining of asteroids, being a peaceful activity, is permissible and lawful.

Apart from peaceful activities, there is a prohibition on issuing "any threat or use of force or any other hostile act or threat of hostile act on the Moon...[and] it is likewise prohibited to use the Moon in order to commit any such act or to engage in any such threat in relation to the Earth, the Moon, spacecraft, the personnel of spacecraft or man-made space objects."[64] The agreement, thus, ushers in an era of amity and concord, and secures a peaceful environment for peaceful activities.

Lastly, there is another obligation imposed in enjoyment of freedom which relates to dissemination of information about activities on the moon to ensure safety of orbital movement and non-interference in operations. The agreement requires that "States-Parties shall inform the Secretary General of the United Nations as well as the public and the international scientific community, to the greatest extent feasible and practicable, of their activities concerned with the exploration and use of the Moon. Information on time, location, orbital parameters and duration shall be given in respect of each mission to the Moon, as soon as possible after launching, while information on the results of each mission, including scientific results, shall be furnished upon completion of the mission ...or periodically..."[65]

The above stated provision ordains that information gathered on missions to the Moon and the results of scientific investigations on the celestial bodies are to be disclosed and widely propagated, which means that such information should become part of the public domain and get assimilated into the knowledge fund of humanity. By inference, this should also apply to visits by robots or astronauts to the asteroids, and this puts secrecy and intellectual property rights relating to natural assets discovered on the celestial bodies in jeopardy.[66] The business dilemma is apparent and discernible.

Hints of Permissibility
The Moon Agreement also betrays scattered hints of permissibility for the use of celestial bodies, including asteroids, for peaceful purposes which could easily comprise mining activities. Though the references are not direct, inferences can be drawn by logical deduction, legal casuistry and liberal gloss over the provisions. Despite all this, many of the conclusions remain shrouded in doubt or can be viewed with scepticism. Some hints are discussed in the succeeding paragraphs to buttress the above contention.

Freedom of Scientific Investigation

A freedom guaranteed under the Moon Agreement relates to "scientific investigation on the Moon by all States-Parties without discrimination of any kind, on the basis of equality and in accordance with international law."[67] This further provides that "the States-Parties shall have the right to collect on, and remove from, the Moon, samples of its mineral and other substances."[68] Thus, digging, quarry and excavation on the moon, and by analogy on the asteroids, is permitted but only for small quantities as samples. A large quantum for commercial purposes is in doubt.

Regarding custody and ownership, the agreement allows that "such samples shall remain at the disposal of those States-Parties which caused them to be collected and may be used by them for scientific purposes. States-Parties shall have regard to the desirability of making a portion of such samples available to other interested States-Parties and the international scientific community for scientific investigation."[69] This clause virtually entrusts permanent possession to the state that caused the materials to be brought from the moon or celestial body. However, it is silent on commercial disposal or exchange for consideration.

There are, of course, precedents where excavating countries have brought small quantities of surface crust from the moon for investigation and experiments. Such samples have loftily been declared as assets of mankind and for scientific benefits to humanity at large. But the reality of possessory treatment of moon-rock has belied such pronouncements as propaganda for goodwill. These altruistic statements have later been recanted to claim virtual ownership of the samples.

For example, the US, through its six Apollo landings, brought back 842 pounds of lunar material. NASA has strictly controlled use of the material, and less than 10 percent has ever been experimented on. NASA has further claimed that the lunar samples are "a limited national resource, a future heritage, and that samples may be released only for approved applications in research, education, and public display." The United States government has vigorously prosecuted anyone arraigned to have improperly obtained any such samples.

The Soviet Union also, through three lunar robotic missions, brought back approximately 300 grams of lunar material for scientific experiments. But a part of it was presumably stolen and has been resold by private individuals to unknown buyers. A small portion of the Russian lunar soil sample was exchanged with NASA in return for a certain quantity of US moon soil sample. This was a transaction with consideration from both sides and, thus, smacks of property transfer.

Both instances clearly establish a streak of ownership for their exclusive handling, possession and control; and the power to exchange what was truly their 'trust holding.' Considering these precedents, right or wrong, it can, in a way, be argued that though celestial bodies themselves, as a whole as they exist in nature, are not subject to appropriation, yet detached portions of a celestial body can be made succursal to state or private ownership and considered as an exclusive resource if so removed from the space body and transported to the earth. An interpretation has been proffered that if an asteroid is removed from its natural orbit by any means, it no longer remains that "original" celestial body and can be prone to exploitation thereafter.

State Responsibility for National Activities
State responsibility is a well established principle of traditional international law and based on this doctrine, similar provisions enunciating state responsibility for national activities in contravention of the law are contained in the Outer Space Treaty and Moon Agreement. Both instruments concede activities on the moon and other celestial bodies by non-governmental entities which can inferentially include individuals and other juridical persons. The only rider that these treaties impose is that such activities shall be carried out under the authorisation of the state and its supervisory control. This mandate seems solely for good governance and linkage to the liability.

The Moon Agreement articulates that states "shall bear international responsibility for national activities on the Moon [and other celestial bodies], whether such activities are carried on by governmental agencies or by non-governmental entities, and for assuring that national activities are carried out in conformity with the provisions set forth in this Agreement. States-Parties shall ensure that non-governmental entities under their jurisdiction shall engage in activities on the Moon only under the authority and continuing supervision of the appropriate States-Party."[70] In fact, responsibility and liability have a symbiotic relationship. Therefore, states are liable *in delictum* of operative norms of activities immaterial of whether committed by state agencies, or non-governmental entities or individual entrepreneurs.

Thus, this provision affords a hint that activities on the moon and celestial bodies, including asteroids, can be undertaken by non-governmental entities which could comfortably connote corporate sector and individual entrepreneurs as *persona juris*. There is no doubt that such activities have to be peaceful which can cover mining of asteroids. But the problem arises whether these entities can remove materials from such bodies in commercial quantities and dispose of them for private consideration and also retain the accrued profits in the personal kitty. This, however, is likely to contravene the principle of common heritage of mankind and its tenets.

Another misnomer relates to the rights to geo-synchronous slots. Stationary orbit positions for television and communication satellites, for instance, are allocated by the International Telecommunications Union. Strictly speaking, they are not "owned" by the assignee, and can be renewed on a regular basis for actual usage and can be leased to other parties for reasons of desuetude. This activity and the Outer Space Treaty's recognition of jurisdiction and property rights for satellites are the basis for a turnover of more than $300 billion per year for the private satellite industry.

An Appraisal

The canvas of present-day space activities shows that the private sector and industry support to governmental space agencies for cargo and crew transportation services by space carriers like Dragon X and Cygnus Spacecraft is bound to stay and progressively expand. Further, public budgets are not the right source for such commercial ventures and outlays for these must flow from the private kitty. Similarly, the corporate sector's contribution to peaceful space ventures like communications, television broadcasting and remote sensing *et al*, whether from the earth or in outer space, is of great consequence and set to prosper.

Further, public budgets are not the right source for such commercial ventures and outlays for these must flow from the private kitty. This trend has entrenched itself for its usefulness and utility and cannot be reversed. However, a clarification is still due on the legality and legitimacy of actions of private enterprise in outer space. Indeed, our specific query here relates to the mining of asteroids, and as a corollary, other celestial bodies, by corporate business entities for commercial purposes.

The Existing Legal Provisions

The Outer Space Treaty

The Outer Space Treaty has no direct and categorical answer and to wade through the complex provisions to hunt for indirect hints of permissibility through the legal labyrinths seems dubious and suspect. It is an honourable treaty between states and binds them to its contents for adherence. *Pacta sunt servanda* is a *mantra* of international law. Thus, private legal entities and nationals are not *sui juris* or true subjects of this *specialibus* international law. Therefore, an inferential deduction will not bestow substantive rights on such entities.

An isolated and ancillary reference in relation to state responsibility and international liability has been liberally glossed to suit vested interests. By object and intent, governmental supervision could allude to certification of space-worthiness of industrial support and products from the earth and documentation of astronauts as nationals. This is, thus, a peripheral provision with no direct mandate or explicit permissibility. Further, the import and logic of the dictum that what is not prohibited is permitted is rather slender. The drawing of adverse inferences from the silence of statutes does not make for robust laws. Therefore, involuted interpretation of space jurisprudence should be unwelcome; it can be unhealthy, even dangerous.

The Moon Agreement

The predicament is no better under the Moon Agreement except that it admits of possible benefits to humanity from the exploitation of material deposits on the moon and other celestial bodies. In view of the limitations of technology at that point in time and the apparent remote possibility of its acquisition and exploitation by the private sector, the agreement is reticent and carries few perspicuous clarifications on the subject. It is, of course, a well-accepted generality that law trails technology and this agreement has followed suit.

The agreement has, thus, not taken a proactive approach of a confident visionary to establish a futuristic legal regime. Hence, a precise answer to our hypothesis is lacking and the solution must lie elsewhere. Fortunately, the Moon Agreement carries an inherent provision for review after ten years so as to adapt itself based on empirical lessons or to technological advancement or contemporary needs of humanity. This promises a possibility for the intended change in law though international politics may foul the due process to introduce suitable adjustments and make it rather arduous.

Asteroids: Movable or Immovable Property

Doubts have been expressed about whether the moon and asteroids are immovable bodies in the universe and, hence, comprise real estate. The argument has it that the celestial bodies are all moving in their respective orbits and their operating apogees and perigees are prone to variation due to the interaction of natural gravitational perturbations in the cosmos. Therefore, to term these as immovable property or celestial real estate would be inaccurate nomenclature. At best, these are floating bodies on predetermined routes that tend to vary at regular or sporadic intervals.

Another line of thinking, taking a cue from floating bodies, treats these as moving rocks. Of course, the natural movement of these bodies cannot be denied and has to be conceded. Scholars of this school have also compared asteroids to the icebergs of Antarctica on the benchmark of their movement but the analogy is incomplete and incongruent on both spatiality and functionality approaches. Icebergs are random roving bodies and their movement is highly prone to circumstantial influences of flow and temperature while celestial bodies have almost fixed orbits and invariable body mass.

The prime characteristics of movability, as opposed to immovability, are susceptibility to external control under human intervention, and possibility to change location by human volition. Both criteria can now be amply satisfied. Human activity can now reach the asteroids and carry on with acts of mining and extraction on and below the surface of the celestial body at will and pleasure. Further, the excavated material can be transported away to deplete the mass of the body. This may be considered as movability in parts or instalments and, thus, immune from legal embargoes.

There is another aspect to human control on 'movability' of an asteroid as a whole body. Scientific knowledge has made it theoretically feasible and development of technological applications will soon render it possible. It is now easily imaginable to alter the orbit of an asteroid or move it closer to the earth or the moon or even a designated industrial platform. Human ingenuity has converted asteroids into movable property, resulting in their realty value and costs of transportation of extracted ores becoming ever diminishing and sensitive to price fluctuations.

The Common Heritage of Mankind

However, in a way, the agreement has shown the vision to acknowledge the possibility for exploitation of the mineral resources of the moon and other celestial bodies as and when appropriate technologies are able, and applicable, to render such activities plausible from all angles. It, therefore, introduced the yet untried and untested, doctrine of Common Heritage of Mankind (CHM) for trusteeship control of the celestial bodies to ensure sustainable extraction of mineral resources, governance of mining companies and distribution of accrued profits among legitimate beneficiaries. It is, like conceptual federalism, intended to foster international solidarity through a common right of humanity to the total celestial realty, to be exercised collectively. However laudable, this doctrine remains controversial, enigmatic and an amorphous notion of contemporary international law.

The Doctrine

The doctrine of CHM constitutes an accumulation of the estate of all celestial bodies in outer space for preservation and shared management towards the benefit of mankind. The peculiarity of this concept is that since it designates CHM as collective and community property, it becomes *res communis* with a life-time share and stake of every living human being at all times, including all future generations. At the same time, these islands of territory are neither subject to state sovereignty on the international plane nor invested in the individual as property rights, flowing from anyone's ingenuity of action or effort of discovery. It is neither divisible nor partitionable to any specific demarcated piece of area for any particular person as heir-apparent. The concept, thus, has two dimensions—inter-spatial for realty and inter-temporal for generations. This virtually makes CHM *res nullius* meaning that it is literally owned by nobody yet collectively owned by all.[71]

This seems congruous to the Hindu Law of Co-parcenary relating to joint holding of family property. The meaning of parcener is sharer and its French equivalent is *arconier* and the Latin, *jartitio*. This law is customary in intestate deaths and concerns the descent of lands as ancestral property of inheritance from a common ancestor to two or more co-heirs possessing an equal hereditary title to them. The distinguishing feature of this inheritance is that there is unity of title, interest and possession claimed by descent that constrains parceners from partition by the writ of traditional law itself. Thus, the concept of co-parcenary appears fairly congruous to the common heritage of mankind where individual shares are not marked with defined limits and *vice versa*. Another analogous doctrine is that of *waqf* among the Muslims.

However, the Moon Agreement, in a nutshell, defines the broad parameters of CHM as envisaged in the concept. These are orderly and safe development of natural resources, their rational management and sustainability for future generations,[72] expansion of opportunities for exploitation of mineral deposits and, lastly, an equitable sharing of benefits by all the states of the world and not the cartel of space-farers. The benefits are broadly comparable to enjoying of the usufructs of the estate.

Critical Analysis of the Doctrine

The doctrine of CHM, howsoever laudable, has not found due favour globally. Even signatory states are reluctant to ratify the agreement, resulting in adherence by only a small number. The industrialised space-faring states are wary of its tenets and depositaries have received few accessions despite the UN sponsorship and support to this treaty.[73] The concept is, thus, mired in doubts and controversies and its original appeal has worn off.

Of course, there are reasons for this disenchantment with the agreement and embodiment of the principles of CHM which advocate inclusivity of all states globally, i.e. even non-participants in space efforts and affirmative action of aid. First, the space-faring nations feel that non-space-faring states that have no contribution in the exploration, prospecting and mining activity, and associated fund utilisation or risk-taking, should have no right to, or stake in,such proceeds, with no inclusive sharing of profits.[74] The proposed arrangement of mandatory doles to all states, irrespective of their participation in outer space activities, is unwarranted, iniquitous and injudicious.

Secondly, in pursuance to affirmative action, the disbursals from profit-proceeds should not be identity-based nor based on classifications of non-development. The action-aid ought to be need-based and welfare should hold the primacy of the ultimate objective. It should stimulate and reinforce a system of empowerment and growth rather than seep in as unproductive subsidy to mask inefficiency, and sloth and scandals. Rightfully, inclusivity funds should be invested in drivers of change and development like transfer of technology and training of manpower for sustained benefits with high multiplier effect on the native economy.

Thirdly, the regime of CHM calls for rational and sustained management of spatial areas, not subject to the control of any state, and mineral resources, yet it remains legally nebulous, and an indeterminate governance model. The economics of enterprise seeks a stable environment of legal norms and production parameters. But this regime as yet provides no surety or security of private right to property or grant of mining leases of reasonable duration to break even to regular profits. In view of the uncertainty of legal permissibility of mining, the corporate sector is shying away from investment planning, thus, inhibiting resource development.[75] The dismal scenario has led some scholars like Martin Menter to pronounce that this has caused a *de facto* moratorium on mining activities in outer space.[76]

Fourthly, it has ideological implications. The CHM model seems fairly akin to Marxist socialism where total property is under state control, while in CHM, it could perhaps be under the trusteeship of the United Nations. Thus, management of production resources would be under bureaucratic supervision and the profits of economic activity shall be shared by the states, by and large, according to their needs, and not in proportion to their effort in the process of production. This economic model, with minor variations, has worked in Communist countries in half the world for over half a century.

It was a great populist idea for revolution but it could not provide a growth-oriented economy because wealth cannot be multiplied by dividing it. Further, as is wisely said, what one person receives without working for, another person must work for without receiving, because the government cannot give to anybody anything it does not first take from somebody else. Disincentive to work and produce is inbuilt in the system that gradually leads to a perception of inequity and injustice. No wonder, in the end, the system has failed in its espoused objective and has collapsed. The Communist economies are faltering and are striving for an overlay of capitalism. Their economic survival is at stake unless they modify their economic ideology.

This statement holds no brief for capitalism for that also needs amends and needs to be revisited with a new mindset. This is only to caution that history does not repeat itself, it is we who do not learn from our mistakes. The CHM of the Moon Agreement must incorporate the drivers of the capitalist economy for optimum wealth generation blended with modified and moderated sharing of proceeds that allocate due rewards to human labour and visible justice to entrepreneurial skills, while, at the same time, showing ample regard to the needs of developing states, in cash and kind.

Lastly, there is an apprehension from the procedural angle. The CHM as enunciated in the agreement is in a raw form, with mere statement of objectives. It has yet to be developed into a proper regulatory framework with statutory authority for enforcement and compliance. International negotiations comprise an arduous and laborious task, beset with tantrums and setbacks that have to be handled with deft diplomacy and pertinacious resilience. This is no easy task and the generic obstacles encountered in the long drawn negotiations over the Law of the Sea Treaty[77] and the hiccups during the conclusion of the Wellington Treaty on Antarctica lead to empirical despondence. Nevertheless, the agreement shall have to be refurbished and the paradigm of CHM evolved with care, and soon enough. Human ingenuity promises to tackle any obstruction.

Need for Review of the Moon Agreement

Another farsighted provision contained in the Moon Agreement relates to its review after a period of ten years from the date it comes into force. It vouches for a futuristic and proactive approach though such a review is optional and need-based on experience gained on its operation. The relevant clause of the agreement reads, "Ten years after the entry into force of this Agreement, the question of the review of this Agreement shall be included in the provisional agenda of the General Assembly of the United Nations in order to consider, in the light of the past application of the Agreement, whether it requires revision."[78] Refurbishing of the agreement is due by the time factor, advance of technology and force of circumstances.

The Moon Agreement can boast of an inbuilt flexibility rarely found in international treaties and similar instruments. This character is revealed by the next clause that states, "However, at any time after the Agreement has been in force for five years, the Secretary General of the United Nations, as depositary, shall, at the request of one-third of the States-Parties to the Agreement and with the concurrence of the majority of the States-Parties, convene a conference of the States-Parties to review this Agreement."[79] It, thus, infuses the possibility of upgrading and sustaining its relevance to times based on lessons learnt from the experience of its effectiveness. Further, it also provides for an easy and simple procedure to requisition a conference when so deemed necessary.

The last clause of this Article is still more radical and forward looking. It mandates a task and purpose for the review conference rather than groping for an objective or merely acting on the agenda of the requisitioning states. The text dictates, "A review conference shall also consider the question of the implementation of the provision of Article 11, paragraph 5 [a legal regime to govern the exploitation of the natural resources], on the basis of the principle referred to in paragraph 1 of that Article [the Common Heritage of Mankind] and taking in account in particular any relevant technological developments."[80] This is also in pursuance to the undertaking given by the states "to establish an international regime...to govern the exploitation of the natural resources of the Moon...This provision shall be implemented in accordance with Article 18[for review of the Agreement] of this Agreement."[81]

Another germane clause provides ample support in tandem to the above provision and elaborates on the procedural formalities to facilitate and formulate an international regime in deference to their undertaking to do so. The agreement urges states, "In order to facilitate the establishment of the international regime referred to in paragraph 5 of this Article, States-Parties shall inform the Secretary-General of the United Nations as well as the public and the international scientific community, to the greatest extent feasible and practicable, of any natural resources they may discover on the Moon."[82] This declaration of discovery assumes significance because all exploitable assets on the celestial bodies merge into the property pool of common heritage while exercise of national sovereignty and private property rights in outer space are strictly prohibited.

These provisions are a great solace and leave sufficient room for specifically devising a governing modality for regulating use and appropriation of locked-up resources on the celestial bodies. It also provides guidelines and principles to work on an updated regime responsive to contemporary human needs on the earth, in concordance with state-of-the-art technology and infusing a comaraderie of public-private partnership because states may not venture into mining projects with funds from the public exchequer. A ray of hope is visible from here and this path can be trodden with confidence to reach legality of mining the asteroids by the private corporate sector. The legal regime, yet to be devised, can be adapted to grant of leasing rights for mining to private ventures for a reasonable period to be sufficiently profitable for sharing dividends.[83]

Need to Establish a Strong Governing Authority

Assuming that desired changes in treaty law can be brought about by international cooperation to enable private enterprises to undertake mining of asteroids and other celestial bodies with express legitimacy, another perplexity accosts us. This relates to the need for a regulatory authority because the relatively low key set-up of OOSA does not appear equal to the task. Perhaps, a World Space Authority (WSA) can be suggested as an alternative to the existing dispensation. The authority could, thus, be entrusted with an enlarged job structure of executive, regulatory and quasi-judicial effectuations.

This organisation in the form and style of a World Space Authority (WSA), under the aegis of the United Nations would implement the legal and regulatory framework for the management of human activity within outer space and on the celestial bodies. The authority so proposed would, therefore, need a proper mandate, a commensurate structure, judicial jurisdiction, the requisite teeth, standard procedures and administrative funding. It could work on the lines of International Civil Aviation Organisation (ICAO) and exercise authority commensurate to the tasks assigned, including grant and management of commercial leases for regulated mining on the celestial bodies. It may also have to discharge duties for inspection and verification to stem overexploitation and encroachments. Some of the functions can be outsourced or delegated to commissions or committees.

An additional task of this authority would be to govern the common heritage of mankind as the total aggregate of the estate of humanity on a trusteeship basis and ensure optimal and sustainable use of celestial deposits and regulate their exploitation through allocation of leases on rent. This system can be likened to the custom of co-parcenery under the Hindu Law for management of joint family estates and a comparable system of *waqf* properties under the Muslim law. The functions of the authority would also entail finalisation of lease agreements on the best terms, monitoring of the operation of commercial leases on celestial bodies, ensuring sustained exploitation to allay environmental concerns, collection of rent from the lessees, a mechanism for conflict resolution, and modalities for distribution of proceeds to the stipulated beneficiaries, as per formulated proportions. This task would demand high standards of probity, impartiality par excellence and stewardship ethics of the highest order.

It is pertinent to mention in connection with grant of mining leases that no company or commercial entity has a fundamental right to a particular lease on specific terms. In a suit on the matter, the Supreme Court of India has judicially decided on merit that no private company has an unalienable right to gain a mining lease or prospecting licence of a designated area. In such cases, discretion exercised may not suffer from any legal flaw or impairment of obligation.[84] Similarly, certain areas on the celestial bodies may, conjecturally, be reserved in public interest for infrastructural needs or for environmental conservation or scientific exploration. The WSA as the leasing authority should have adequate powers for such compulsive reservations in the larger interest of mankind.

There is, however, a rebellious view that attempts to demolish the above arguments. It maintains that the very idea that everything in the universe is somehow collectively owned by the governments and that it is up to them to grant 'permission' to everyone else to use anything is obnoxious and pernicious. This is an utterly corrupt, inverted view of property rights. The sane laws of the earth uphold that rights flow from individuals and even for unclaimed lands, the first right on sighting goes to the individual and then only gets transferred to the state by the nexus of nationality.

In the same vein, those pioneer corporations investing in Research and Development (R&D) and working to gain access to an asteroid and mining its natural resources should be able to, by all logic, own and appropriate this property. It is well known that mining of an asteroid is an extremely complex, difficult, life-risky, and expensive endeavour; the fruits of such labour cannot be denied and those who undertake such an endeavour, should, under all arguments, own the results of their risk-taking. The state can legitimately tax the income or the gains but cannot snatch away their fundamental economic right to the property. On the contrary, the sole rational purpose of a government is to uphold and protect the rights of its nationals, including, critically, their property, irrespective of its location. That is a duty of the state that it cannot abdicate. In the worst eventuality, if the asteroid miners are sensible and capable, they will ultimately form their own extra-terrestrial governments in order to safeguard their inherent right to property, including against the looting ambitions of terrestrial governments and earthlings.

Conclusion

The appraisal compels us to conclude that the existing space law affords no scope for legitimately deducing that mining of asteroids by private enterprise for commercial disposal of material so extracted and to be sole retainer of profits so made, is licit and legal. No provision of the Outer Space Treaty or the Moon Agreement permits large scale excavation of ores from celestial bodies for processing or sale *in situ* or on the earth because they have neither mining lease rights to the mineral deposits nor property rights to the extracts under the common heritage of mankind regime. Further, profits from such ventures, if permitted to operate, are to be apportioned equitably with the developing states and a reasonable share also paid to the space-faring nations that helped to make such exploitation technically feasible and economically viable. However, the modalities and proportions for division of proceeds among the beneficiaries are yet to be devised.

A rational conclusion prods us towards liberal grant of property rights on celestial bodies because these can be useful drivers and effective stimuli for development and growth beneficial to mankind for the avowed purpose. Thus, the tussle and choice is between grant of individual property rights or resigning to a slow rate of progress in the realm of outer space and utilisation of resources so vitally required on earth for global uplift. The only rider should be judicious regulation of such rights and mining leases for sustainable exploitation of natural resources. This aspect assumes importance because one can pre-sense a shift in strategic paradigm in which future star wars of the 21st century would be less for military objectives than for control over natural resources of celestial bodies.

In the end, it must be accepted that a new legal regime with effective regulatory mechanisms for commercial exploitation of extra-terrestrial resources needs to be put in place for proper governance of the celestial treasure. This may need an organisation in the form of Celestial Resources Authority (CRA) or World Space Authority (WSA), to work as an organ of the UN albeit with additional tasks and commitment of managing commercial leases on celestial bodies, monitoring of sustained exploitation as per lease terms, quasi-judicial machinery for dispute redressal and distribution of residual profits to the lawful beneficiaries. It is, of course, a proposal in raw skeletal form that can be widely debated and accordingly suitably refined.

Notes

1. Treaty on Principles Governing the Activities of States in the Exploration and Use of Outer Space, including the Moon and Other Celestial Bodies, 1967. (In short, the OST or Outer Space Treaty.)
2. Ibid., Article VI.
3. Agreement Governing the Activities of States on the Moon and other Celestial Bodies, 1979. Popularly called the Moon Treaty or the Moon Agreement.
4. Ibid., Article 8 (1).
5. Accessed on August 8, 2012.
6. Reported in *The Telegraph* (USA). Details yet to be officially announced.
7. *The Daily Mail*, (London), July 21, 2012. http://www.business-standard.com/generalnews/news/asteroid-as-big-ascity-block-to-hurtle-past-earth/35012/. Accessed on July 21, 2012.
8. Two "Mainbelt" asteroids were provisionally discovered in August 2012 by four schoolboys from Delhi (India) which have been confirmed by further observations. The team used a complex procedure called "Astrometrica" using exclusive data to look at a specific part of the sky for moving objects over a period of time. It is a rare feat deserving of commendation. http://zeenews.india.com/news/space/young-delhi-boys-discover-asteroids_791993.html. Accessed on August 8, 2012.
9. L. Wilson, K. Keil, S. J. Love, "The Internal Structures and Densities of Asteroids". *Meteoritics & Planetary Science* 34 (3), 1999, pp. 479-483.
10. Nikku Madhusudhan and Olivier Mousis are astronomers forming an Indo-French team who discovered a new planet.
11. Reported by Chris Wickham, *REUTERS*, First posted: Thursday, October 11, 2012 09:36 AM EDT.
12. For additional information, see John S. Lewis, *Mining the Sky: Untold Riches from the Asteroids, Comets and Planets* (Perseus, 1997).
13. Brian T. O'Leary, "Mining the Apollo and Amor Asteroids, 197," *Science* (1977):363. Also see Tom Gehrel, ed., *Asteroids* (1979).
14. For more details, refer Mark Sonter, *Space Future*. accessed on June 8, 2006.
15. SOPL is a constellation of reflectors, deployed onto orbits, to channel power from the sun to a chosen point in space.
16. (PDF), *Interplanetary Collision Hazards* (Pasadena, California, USA: Jet Propulsion Laboratory, NASA, 1998), accessed on July 1, 2012.
17. Psyche is an asteroid of the M-type with a diameter of 1 km.

18. "Tech Billionaires Bankroll Gold Rush to Mine Asteroids", *Reuters* News Report, April 24, 2012.
19. John Brophy, Fred Culick, Louis Friedman *et al.*, Keck Institute for Space Studies, California Institute of Technology, April 2012.
20. David L. Kuck, "Exploitation of Space Oases", Proceedings of the Twelfth SSI-Princeton Conference, 1995.
21. Robert Freitas, William P. Gilbreath, eds., NASA Conference Publication CP-2255 (N83-15348, 1982)
22. Reasons discussed later in the chapter. Under international legal requirements for establishing a claim to territory or property, the discoverer must touch, or step on, the land body.
23. It was reported in newspapers in 2011 that Ms. Mayawati, the then Chief Minister of UP (India), was gifted one such plot on the moon.
24. Natalie Wolchover, "Life's Little Mysteries", space.com, April 24, 2012.
25. Virgiliu Pop, *Who Owns the Moon? Extra Terrestrial Aspects of Land and Mineral Resources Ownership* (Springer, 2009).
26. Ibid., p. 2.
27. Ibid., p. 3.
28. Ibid., p.4.
29. Ibid., p. 6.
30. Ibid., p. 2.
31. Ibid., p. 5.
32. Nemitz v/s the US (2004), US District Court, Nevada.
33. The Preamble to the Outer Space Treaty.
34. Declaration of Legal Principles Governing the Activities of States in the Exploration and Use of Outer Space, UN General Assembly Resolution 1962 (XVIII), adopted on December 13, 1963.
35. Treaty on Principles Governing the Activities of States in the Exploration and Use of Outer Space, including the Moon and Other Celestial Bodies, 1967, Article I.
36. Ibid., Article II.
37. Ibid., Article I, paragraph 2.
38. Ibid., Article III.
39. Ibid., Article I, paragraph 2.
40. Ibid., Article I, paragraph 3.
41. Ibid., Article IX.
42. Ibid., Article IV, paragraph 2.
43. Ibid., Article IV, paragraph 1
44. Ibid., Article IV, paragraph 2.

45. Treaty on Principles Governing the Activities of States in the Exploration and Use of Outer Space, including the Moon and Other Celestial Bodies, 1967, Article VI.
46. Ibid.
47. *Agreement Governing the Activities of States on the Moon and other Celestial Bodies, 1979.* Popularly called the Moon Treaty or the Moon Agreement. Refer the Preamble.
48. Ibid., Preamble. Words in parentheses are added.
49. Ibid., Preamble.
50. Ibid., Article 1.
51. Ibid.
52. Ibid. Article 1, paragraph 3.
53. Ibid., Article 4, paragraph 1. Words in parentheses are added.
54. Ibid., Article 11, paragraph 1. Words in parentheses added, based on applicability under Article 1.
55. Ibid., Article 11, paragraph 2.
56. Ibid., Article 11, paragraph 3.
57. Ibid., Article 11, paragraph 5.
58. Ibid., Article 11, paragraph 7.
59. Ibid., Article 8, paragraph 1.
60. Ibid., Article 8, paragraph 2.
61. Ibid., Article 8, paragraph 3.
62. Ibid., Article 15, paragraph 1.
63. Ibid., Article 3, paragraph 1.
64. Ibid., Article 3, paragraph 2.
65. Ibid., Article 5, paragraph 1.
66. Ibid., Also refer Article 11, paragraph 6.
67. Ibid., Article 6, paragraph 1.
68. Ibid.,Article 6, paragraph 2.
69. Ibid.
70. Ibid., Article 14, paragraph 1. Words in parenthesis are added.
71. For more details, refer Kemal Baslar, *The Concept of Common Heritage of Mankind,* (Martinus Nijhoff Publishers, 1998), p. 80ff.
72. Ibid.
73. As on March 30, 2012, only 17 states are parties to this agreement. Most spacefaring countries have neither signed nor ratified the agreement. India, though a signatory, has not ratified yet.
74. Ricky J. Lee, *Law and Regulation of Commercial Mining of Minerals in Outer Space* (Springer: Space Regulations Library, 2012), p. 203.
75. Ibid., p. 204.

76. Refer Martin Menter, *Astronautical Law* (US Industrial College of Armed Forces, 1959). Digitised in September 2010.
77. UN Convention on the Law of the Sea.
78. *Agreement Governing the Activities of States on the Moon and other Celestial Bodies, 1979,* Article 18.
79. Ibid.
80. Ibid., Words in parentheses added for clarity.
81. Ibid., Article 11, paragraph 5. Words in parentheses added for clarity.
82. Ibid., Article 11, paragraph 6.
83. Lee, n. 74. He has suggested "to create a new regulatory regime."
84. Monnet Ispat and Energy Ltd, vs. Jharkhand and Central Governments. Supreme Court Cases, July 2012. A gist of the verdict was published in *The Times of India,* July 28, 2012.

7

International Code of Conduct for Activities in Outer Space: An Exercise in Futility

Introduction

Space law regulating human activities in outer space was developed with great enthusiasm by the UN member-states. Their understanding of its importance, appreciation of the stakes and realisation of its necessity displayed remarkable unanimity and rare commonality of purpose and intent. The first two decades of the commencement of space activities were marked with enunciation of fundamental principles for governance of outer space and negotiation of treaties and agreements as substantive law. The corpus of space law grew with the necessity to tackle problems that accosted current operations or presented challenges for future activities or even to promote international cooperation for the safety of astronauts and the security of satellites. The international attitude was so cordial and cooperative that even the principles and guidelines articulated during this period were voluntarily adhered to, with little reservation and least objection. This was an era of prolific growth of space law tailored to contemporary needs and foreseeable contingencies. The *grundnorm* had been laid.

Somehow, for some reason, the evolution of space law has been almost moribund for the last four decades since the conclusion of the Moon Agreement. This too has been no significant achievement considering the low number of accessions to this instrument. However, over this period, a number of principles and guideline documents on specific issues have been promulgated under the aegis of the United Nations but their impact is low and their character non-obligatory. This is hardly complimentary when over the same period, space activities have expanded exponentially and diversified vastly. Space technology has advanced immensely and rather rapidly to explore deep in space and yield opportunities to exploit new vistas of space economics and the celestial natural resources that are in prolific abundance. Almost concurrently, space debris has multiplied to threaten the sustainability of the space medium for human usage for peaceful and welfare activities. A well known scholar laments, "It is surprising that for a realm as critical as this, so little attention has been paid to keep space law apace with the changing times. In fact, in the last couple of decades, as the profile of usage of space has changed drastically, efforts at forming new legal regimes to prevent, or at least regulate, some kinds of activities have had little traction in the international forum."[1] The stakes are common and ominous.

Daunted by the laborious and dilatory procedures for the treaty negotiations and in the wake of the thrust for soft laws and buoyed by the ease in their promulgation, a trend has emerged to introduce principles, guidelines and declarations. In the same vein comes the initiative of the European Union Council to sponsor the International Code of Conduct (ICoC) for Outer Space Activities. A perusal of the contents of the ICoC reveals that it makes no innovative breakthrough nor provides elaboratory explications on grey areas. This is merely reiterative of the provisions existing in the current international instruments on space law. At best, the code issues an appeal to voluntarily adopt recommended best practices, devoid of any binding obligation or sanctions for infringement.

This instrument, thus, traverses no progress to confront the realities created by advanced technology applications e.g. space tourism or mining of celestial bodies or traffic control in space or colonisation of the moon or deep-space probes to explore geo-chemistry or amplification on the common heritage of mankind *et al*. The new breed of problems is real and can be grappled with, by innovative thinking of cooperative minds and not by repeated rhetoric of platitudes. Thus, the futility of such an exercise is vividly obvious and reflective of timidity and helplessness to embark on treaty negotiations to tackle contemporary realities. Viable solutions have to be found and procrastination, of course, is never the best answer. This chapter, therefore, makes an honest appraisal of the possible advantages accruing from the ICoC vis a vis the perceived necessity, or lack of it, for such an exercise in diplomacy or the very instrument itself.

The International Code Of Conduct

The Evolution

For two decades following the pioneering space forays in the Sixties of the last century, there was hectic activity and intense negotiations to determine the regulatory provisions for the continuing man-caused intrusions in outer space. Legal luminaries actively participated to establish the fundamentals of space law. These two decades were really fortuitous and yielded a crop of space law of five international instruments comprising binding treaties, a convention and agreements that set the tone for efficient and cooperative space relations. There were some grey areas and loose ends but these were honestly acknowledged and solutions promised in the future with the unfolding of the direction of human effort and development of space technology.

The United Nations has been a key player in the encouragement of international cooperation in the space segment and has, thus, played a sterling role in the regulation of space activity since the dawn of the space age. Its first tangible step in this direction was the constitution of the UN Committee on the Peaceful Uses of Outer Space (COPUOS) in 1961. This committee has done commendable service in debating and propagating some unconventional principles applicable to outer space. It has been central to the drafting, formulation and facilitation of international treaties which form the basis of the international law of outer space. The committee continues in its unfinished task, though the pace of tangible achievements has slowed down considerably.

The following period of three decades saw virtual inactivity in space law formulation and a lull in the deliberations on substantive space issues. The legal inactivity and stagnation was unfortunate and palpable and though there were technology happenings of import and impact, these were disregarded and solutions procrastinated. As is wisely said, nothing happens in the waiting room except the passage of time. Sometimes, it may afford stimulus for cogitation or hold hope for an opportunity. But, this failed to happen; time and opportunity both were persistently lost.

Thus, in the face of newer developments in technology for scientific applications and human determination for riskier ventures, stable legal solutions were not attempted. Just to cope with some urgent situations and persistent eventualities, scholars resorted to normative law in terms of principles and guidelines. The response was not equal to the task nor direct enough. These soft law instruments had endemic frailties in that these were neither efficacious formally nor binding on contingencies.

From the same crop came a *suo motu* proposal from the European Union for a Code of Conduct for Outer Space Activities ostensibly triggered by the Chinese Anti-Satellite (ASAT) test in outer space in February 2007 though this was quickly followed by the US demonstration of similar prowess in January 2008. The evolution of the code started in the autumn of 2007 during the Portuguese EU Presidency. This draft code was later updated and converted into Best Practice Guidelines by the successor Slovenian Presidency of the EU. Informal consultations took place within the EU and with key space-faring countries like the US, Russia and China.

The following Presidency of the EU by France actively pursued this initiative and a Draft Code of Conduct was officially presented to the Council of the EU as an instrument of a complementary mechanism to the existing framework governing outer space activities. The draft was circulated to member states in December 2008 and was widely debated and discussed. As a result of these deliberations, certain amendments were proposed which have been incorporated to make it truly representative of common intent and, hence, more acceptable to the EU nations.[2]

Later, considering the wider appeal of the subject and universal applicability to all states, the EU revised its opinion and has now globally floated this modified draft in September 2010. It was also retitled as an International Code of Conduct (ICoC) for Outer Space Activities and presented for international consultations with third countries and other like-minded non-space-faring states. It was also opened for participation in multilateral negotiations by Regional Integration Organisations and International Intergovernmental Organisations for eventual endorsement to the code.[3]

The ICoC has been in the public domain for some years but has generated rather limited public discussion and attracted still less attention beyond the European Union states. The final text was to be opened for signatures by all countries and permitted organisations in 2012 but owing to certain complexities has now been postponed till a later date. This version may be further revised to accommodate diverse viewpoints. For the present, it has been referred to the UN Institute of Disarmament Reseach (UNIDIR) for an outreach programme to mobilise opinion and build a consensus on the code.

To reiterate, "This code essentially seeks consensus on an idea of voluntary and nonbinding best practices and transparency and confidence-building measures in regard to various activities in outer space. The code expects the signatory states to declare their ongoing and proposed activities in space."[4]

> The European Union officially launched...in Vienna the multilateral diplomatic process to discuss and negotiate its initiative for an International Code of Conduct for Outer Space Activities...110 participants from more than 40 countries gathered for this multilateral meeting, at which the European Union introduced a revised version of its Code, based on comments received in bilateral meetings with various partners. Substantial negotiations on the basis of this text will start at the Multilateral Experts Meeting of October 2012, which will be open to the participation of all UN Member States, with a view to adopt the Code in 2013.[5]

The proposed International Code, on adoption, is expected to be applicable to all outer space activities conducted by states or non-governmental entities, and would lay down the basic rules to be observed by space-faring nations in both civil and defence space activities. This initiative is already supported by a considerable number of space-faring nations, including among them the US, India and Japan. Further, the UNIDIR has also officially launched a parallel project with a diplomatic process to facilitate information, documentation and exchange of views on the concept of the proposed code.[6] Of course, finality is yet to be reached.

Motivations for the Code

The motivation for the code is important for our appraisal and, thus, deserves its rightful place and discussion here. It can be identified from the Preamble to the code and an adumbration of its objects and purposes is attempted in the succeeding paragraphs.

First, the code recognises that "space activities are expanding and their importance is crucial. Space is a resource for all countries in the world, and those which do not yet have space activities will have them in the future… and believes a pragmatic and incremental process can assist in achieving this goal…[Thus, ICoC is] a means to achieve enhanced safety and security in outer space through the development and implementation of transparency and confidence-building measures."[7]

Secondly, the code considers that space "activities play a growing role in the economic, social and cultural development of nations, preservation of the environment, promotion of international cooperation, strengthening of national security and sustaining international peace." It further deems "that space capabilities – including associated ground and space segments and supporting links – are vital to national security and to the maintenance of international peace and security."[8]

Thirdly, the code "aim[s] at promoting a peaceful, safe and secure outer space environment…" in the face of growing "[S]pace debris [that] constitutes a threat to outer space activities and potentially limits the effective deployment and exploitation of associated space capabilities." It, therefore, stresses "that the growing use of outer space increases the need for greater transparency and better information exchange among all actors conducting outer space activities;" and is "[c]onvinced that the formation of a set of best practices aimed at ensuring security in outer space could become a useful complement to international space law…"[9]

In a nutshell, the code recognises "that a comprehensive approach to safety and security in outer space should be guided by the following principles: (i) freedom of access to space for peaceful purposes; (ii) preservation of the security and integrity of space objects in orbit; (iii) due consideration for the legitimate defence interests of states…" And is "[c]onscious that a comprehensive code…could contribute to promoting common and precise understandings..."[10] Of course, "The present Code is applicable to all outer space activities conducted by a Subscribing State [or jointly with other states or by entities within its jurisdiction.]" and "[A]dherence to this Code and to the measures contained in it is voluntary and open to all states."[11]

Other Versions

The United States
On the lines of the EU Code, the US is also contemplating sponsoring a code from its own perspective and embodying its own national interests *albeit* couched in polite terms. In fact, the Henry L. Stimson Centre of Washington University has already taken an initiative and drafted a Model Code of Conduct (MCoC) which is in circulation and being keenly debated. This code appears more fundamental, less controversial and broader in spectrum. It seems well ordered on subjects, better worded and more expressive of content. Nevertheless, this also suffers from the malady of repetition of provisions contained in the existing corpus of space law: whether treaty law or normative law. It does not break new ground nor is it innovative on substantive gaps or amplificatory on grey areas of contemporary law.

At the same time, NASA is also working independently on a similar Code of Conduct for Outer Space though it has come up with no major differences with the other codes in circulation. In all probability, it may not be substantially different from the European Code or the Model Code except that this would reflect the US official governmental perspective and be expressive of its overt consent. Its appeal to the world at large is uncertain till promulgated and scrutinised in detail. But its boldness, if at all, would be limited to the US self-interest and continued hegemony. This is understandable unless it somehow turns otherwise objective and altruistic to place global and cosmic interests before vested national compulsions and, thus, become an international instrument of wide solicitation and durable value.

It is also pertinent to mention that the US has given overt endorsement to the cause of disarmament and arms control in outer space as a universally commendable objective but is rather reticent to accord formal assent to any such doctrine for fear of bondage. The US is, therefore, extremely cautious in primordially protecting its interests and aggressive in progressively promoting its committed security concerns, including those of its allies; and vested stakes conducive to its national priorities.

The Chinese Variant

China considers its perspective unique and deserving of a separate code for its conduct in outer space. Unfortunately, China has indulged in aberrative space behaviour, which, though not technically and legally infractious in view of ostensible reasons yet appears less pragmatic and more damaging to the space environment. Such delinquent activities cannot be justified by codes, irrespective of the origin or the author. Outer space is common heritage and its sustainability for common use as well as bequeathing it in good shape for the future generations is also our prime responsibility and a fiduciary obligation as present trustees.

Basically, the Chinese objection to the ICoC relates to the requirement for an open declaration of the space policy of states. This is hardly serious enough to merit reconsideration. And further, China considers itself as the "established Asian hegemon in space capabilities" and has a grouse that it was not part of the creative process from the beginning to present its perspective. This stance refuses to accept that the code was initially an in-house drafting of norms and in this background, the EU could not offer this opportunity to third parties to participate in deliberations and present their stand. The Chinese reservation on this issue seems trivial and stems purely from the pride of their burgeoning economy, geo-political weight, threshold of space capabilities and diplomatic niche.

China and Russia, in collaboration, are becoming "increasingly vocal proponents for negotiating on Prevention of an Arms Race in Outer Space (PAROS) in the Convention on Disarmament (CD). This agenda item gained near universal support in annual UN General Assembly Resolutions..."[12] Of course, the emphasis here is on strategic security and prohibition of weapons in space rather than debris mitigation. The shift in focus may be intentional and deliberate. The US has vehemently objected to this and dubbed it as a tactic of the Cold War era. Perhaps, it could be a propaganda ploy to gain global attention and underrate the intrinsic appeal of the proposed EU code.

However, it is important to take notice that the same partners[13] had in 2008, "introduced a draft treaty that would extend the OST's ban on WMD in space to prohibit all types of orbiting weapons. It would also explicitly ban the threat or use of all types of force against space objects."[14] The proposal is titled "Treaty on the Prevention of the Placement of Weapons in Outer Space, Threat or Use of Force against Space Objects." (PPWT). Despite the desired emphasis on space free of weapons, this proposal is woefully silent on the modalities for verification of compliances. Thus, no doubt, this concept highlights security concerns of validity and importance, yet is weak in verifiability. The US actually dismisses this offering "…because it would outlaw US deployment of space-based missile defense interceptors" but makes an overt pretext that it would be "impossible to define and verify a ban on space weapons without impeding the peaceful use of space."[15]

The Asian Approach
Some of the Asian nations like Japan, South Korea, Malaysia, North Korea, Singapore *et cl* are also growing up in space affairs and should participate in such formulations proactively and express their viewpoint eloquently to protect their genuine and nascent interests rather than acquiesce belatedly. The European effort appears incomplete in scope and repeats the obvious; and, thus, falls short of its own objectives. It needs to be comprehensive and should not restrict itself to enunciating solutions to contemporary concerns. It should be futuristic and proactive to encompass other ensuing issues like proliferating nuclear devices and power plants in outer space or elaboration of rules relating to commercial mining in space and the common heritage of mankind and other matters that seek amicable resolution and clamour for urgent attention.

It is from this mark that Indian and Asian space-farers can ambitiously embark to draft a framework called the Asian Code of Conduct that should be a norm-shaper. This proposal can take off following a different approach and perspective to address common imperatives while still retaining its popular appeal and universal acceptance. It may be suggested that this initiative must also involve Australia in a genuine dialogue to gain better clout and attention. Nevertheless, multiplicity of codes can be harmful to the cause and an alternative could be to participate in the ICoC and press for the inclusion of the respective perspectives or the desired emphasis on select points.

From Asian prioritisation of major issues concerning outer space, the escalating arms race and not space debris assumes greater importance. Whatever be their common views regarding holistic security, space traffic management or modalities of global governance or reservation about the publication of space policy or compulsion to enshrine rules and procedures in national legislations, should find the correct iteration and right place in their draft document. This would be the right attitude rather than staying aloof on egotist credentials.[16]

The Indian Stand

India is also being coaxed by strategists to firm up its approach and perspective either as a separate proposal or be vocal of its specific stand in multilateral negotiations on the EU ICoC. One really wonders, when the ICoC hardly deviates from the existing documented instrumentalities on space law, where is the scope for India to differ or strive for its specifics? In any case, it is an inadequate conceptual framework for space governance and global security. It also lacks a holistic approach to space activities and is segmental in its focus. Nevertheless, however trivial, one needs to be sure of the conflictive points that India harbours and which require consideration, compromise or resolution.

It seems that India ostensibly laments the lack of a institutionalised enforcement mechanism and absence of a legally binding provision for compliance verifications. This is a rather sensitive issue and impinges on sovereignty. It is surmisable that in its turn, India may also be wary of such overt openness and revealing exposure to verificatory and intrusive inspections. And India has a further grouse that it was not part of the creative process from the beginning and has not contributed its perspective.

Whatever be the background to offering or denying this opportunity, the reservation seems trivial and stems purely from the ego of its self-assumed strategic importance, confirmed space capabilities, geo-political lobby or as a sheer diplomatic formality.

Nevertheless, some strategists hold that for India it is a no loss situation of implicit acceptance because no established ideological position is compromised and yet it need not fall foul with friends and allies. Accession to such an innocuous and unobtrusive regime neither places any serious embargoes nor does it formally bind to compliances. Without being hypocritical or irreverent, the option seems of no serious consequence.[17]

The Existing Corpus of Space Law

The UN has made a sterling contribution in the evolution of space law and the entire world has cooperated and displayed singular determination to support in unison the UN Resolutions regulating human ventures in outer space. These resolutions lay down the principles governing activities in space, some of which are at wide variance from the traditional international law, yet the acceptance has been near unanimous. This followed a period of great cooperation during which many treaties, agreement and guidelines were negotiated with great ease and speedily concluded. It was the season of the evolution of space law in treaties and agreements. Soft law in the form of principles and guidelines was also adopted by the UN General Assembly in the later years.

Space Treaty, Conventions and Agreements

A few illustrations will vindicate the above statement. Apart from drafting the UN Resolutions, the most significant work of COPUOS has been the Outer Space Treaty.[18] This required great deliberation because outer space is an extraordinary medium in many respects and demanded unique legal principles that would facilitate international relations and govern human activities in outer space. This treaty, thus, established the crucial understanding that outer space and the celestial bodies are the common heritage of mankind for sharing benefits, and under the concept of global commons, these are not subject to the territoriality principle of national appropriation and sovereignty; there is freedom of activity in space without frontiers, and the states are responsible for their own activities and those of their nationals and bear an obligation not to despoil or harm the outer space environment. These were relatively new concepts contrary to established precepts of international law but their acceptance was without resistance or any objection.

Other treaties and agreements facilitated by COPUOS that deserve mention are: the Agreement on the Rescue of Astronauts, the Return of Astronauts and the Return of Objects Launched into Outer Space. This was adopted on December 19, 1967, and came into force on December 3, 1968. It admits of national jurisdiction over astronauts and space objects; the Convention on International Liability for Damage Caused by Space Objects was concluded on November 29, 1971, and became effective from September 1, 1972; the Convention on Registration of Objects Launched into Outer Space was adopted on November 12, 1974, and entered into force on September 15, 1976. Another major instrument was the Agreement Governing the Activities of States on the Moon and other Celestial Bodies that was adopted on December 5, 1979. This agreement came into force with effect from July 11, 1984, but it still cannot boast of many adherents.

Allied Treaties Regulating Outer Space Activities

It is equally pertinent to recognise the value of mechanisms currently in place which, though not directly related or exclusively negotiated on space issues, yet are germane in regulation or even applied to governance of outer space activities. For example, the Constitution and Convention of the International Telecommunication Union, (ITU) 1994 and its Radio Regulations (1995), as amended, that have facilitated distribution of the radio spectrum and allocation of geo-stationery equatorial slots. The service provided and control exercised by ITU is commendable, indeed, exemplary.

There are several more treaties and regulatory regimes pertaining to the nuclear test ban and arms limitation which though not reflecting immediacy, yet have a proximate bearing on the health of the space environment and safety of operations in outer space. Some of these are mentioned below but this enumeration is by no means complete or comprehensive.

- The Treaty Banning Nuclear Weapon Tests in the Atmosphere, in Outer Space and Under Water, 1963, popularly called the Partial Test Ban Treaty. It prohibits nuclear explosions everywhere, whether for military or peaceful purposes, except underground.
- The Comprehensive Nuclear Test Ban Treaty, 1996.
- Intermediate Range Nuclear Forces Treaty, 1988.
- Strategic Arms Limitation Treaty, 1994.
- The International Code of Conduct Against Ballistic Missiles Proliferation, 2002.
- Treaty on Strategic Offensive Reductions, 2003.[19]

The United Nations Charter

In addition to the above, one must remember the grand provisions of the Charter of the United Nations and the application of international law to the activities in the exploration and use of outer space, including the moon and other celestial bodies.[20] The salience of the following Article of the UN Charter deserves reiteration for better appreciation.
- Article 2(4) obliges all members to refrain from the threat or use of force against the territorial integrity of any state. Though state sovereignty does not exist in outer space, including the moon and celestial bodies, yet national jurisdiction in space assets, whether in outer space or on the earth, is recognised by space law.
- Article 42 authorises that the UN Security Council may mandate collective action by members involving use of force, as may be necessary, to maintain or restore international peace and security.
- Article 51 recognises the inherent right of self-defence of all states.

Principles, Guidelines and Declaration

Apart from the above, a lot of soft law principles and guidelines have been recommended by the United Nations through General Assembly Resolutions. These may not, thus, appear mandatory for compliance yet their binding force is rather high in operational and functional conditions. State practice accords these high credence and adherence. The very nature of these instruments highlights the importance of international cooperation

and the emphasis put on these by states volitionally confirms the same. These are listed below:
- 1. Principles Governing the Use by States of Artificial Earth Satellites for International Direct Television Broadcasting, 1982.
- 2. Principles Regarding Remote Sensing of Earth from Outer Space, 1986.
- 3. Principles Relevant to the Use of Nuclear Power Sources in Outer Space, 1992.
- 4. Declaration on International Cooperation in the Exploration and Use of Outer Space for the Benefit and in the Interest of All States, Taking into Particular Account the Needs of Developing Countries, 1996.
- 5. The Recommendations on the Practice of States and International Organisations in Registering Space Objects, 2007 (as contained in UN General Assembly Resolution 62/101).
- 6. Guidelines for Mitigation of Debris in Outer Space, 2007. These have been developed by the Scientific and Technical Sub-Committee of the Committee on Peaceful Uses of Outer Space, (COPOUS).

A Favourable Appraisal

It appears that an earnest effort at global cooperation in space activities emanates from the European Union's initiative of the International Code of Conduct for Outer Space Activities[21] that encourages cooperation, consultation and mutual assistance to "seek solutions based on an equitable balance of interests." This accepts to redefine and impart a new understanding of sovereignty that is different from the one imbedded in traditional international law. The EU has urged nations worldwide to join and adhere to this code of a set of best practices to make its acceptance almost universal.

The ICoC expresses apprehensive concerns about the ever increasing space debris in order to highlight its risks for the safety of space operations and the need for mitigation for avoidance of collision hazards and, thus, maintain sustainability of space as a medium for future activities. It also alludes to growing weaponisation of outer space and seeks codification of best practices for transparency and confidence-building measures without disturbing the inherent right of the states for self-defence under the provisions of the UN Charter. It also exhorts for adherence to the essential treaties like the OST, Rescue Agreement and Convention on International Telecommunications and Radio Regulations. Lastly, it urges members to formulate and publicise their national space policy and procedures in tandem to maximise security, minimise accidents and incorporate consultative resolution of disputes based on "equitable balance of interests."

The impelling motivation for this code stems from the apparent success of a few parallels in the recent past. The first example that could be cited pertains to wide acceptance of the Hague Code of Conduct against Ballistic Missiles Proliferation, 2002 (H-COC). This code pertains to the arms control mechanism on proliferation, established to induct transparency in the missile domain. The code expected the signatories to honour the information sharing regime but the expectations were belied. Success was only partial and hardly elevating. Further comparability is that this instrument is voluntary in adherence and non-binding in controls. Interestingly, this H-Code has been signed by 148 countries but the irony lies in the fact that "most of the signatories to this mechanism have no missile capabilities."[22] It may also seem pertinent to mention that in the space segment, leave aside a voluntary code, regrettably, a binding instrument like the Convention on Registration of Space Objects has not been fully respected by some states even after ratification.

Other examples could be the Biological Weapons and Chemical Weapons Conventions. The Biological Weapons Convention is good in substantive part but is deficient in providing an effective mechanism for verification of compliances stipulated under the convention. Thus, its delinquency is common to the breed of such international instruments and this particular case remains with imperfections, to survive with limited success. However, the Chemical Weapons Convention has been rated better in most aspects and has often "been touted as a perfect treaty,...[but, unfortunately, it suffers] from betrayal by the United States and Russia, both of which are far from fulfilling their commitment of total destruction of chemical weapons from their inventory."[23] The comparison here too appears incongruous considering the importance of the stakes and the appeal of the subject.

Another subaltern principle advocated by the code is to promote confidence-building measures and transparency in outer space activities for common and precise understanding.[24] The concept is, indeed, laudable but the code suggests no concrete measures or parameters of transparency. One such method culled out from the code makes it incumbent on the subscribing states to place their space policy and ongoing activities in the public domain.[25] But many countries are reluctant to openly promulgate their policy and long-term objectives, resulting in hedging strategies to make incumbency both invalid and suspect, or encouraging hypocrisy and clandestine missions. And as a corollary, the code expects that the states announce vehicle launches and satellite missions in due time, preferably in advance.[26] The idea, however admirable, lacks a legal obligation for compulsive adherence. Therefore, it may not be fully complied with; it has been suspected that China has defaulted in the past in the requirement of registration or has resorted to it rather late. Hence, the futility of this proposal becomes obvious because if a mandate has been defied with impunity, the code will be treated still worse.

A Critical Appraisal
Let's now embark upon a critical analysis of the objects, the contents, the utility and the basic merit of the ICoC and examine the advantages likely to flow from such an international instrument vis-a-vis the effort involved in negotiations and other formalities connected therewith. Further, there is a risk of a spate of codes by other nations putting forward their viewpoint from their individual national perspective. A couple of these are already in the offing, namely, the US Code and the Chinese Code (as the PPWT) or the Asian Code. This proliferation, with varying emphasis or highlighting different lacunae, will tend to be confusing and may cause polarisation with divided loyalties. From here, it can be surmised that multiplicity of codes with different slants to, and variant individualised emphasis on, established law, may lead to near-anarchic public order in outer space. This can hardly be permitted.

Repetition of Existing Provisions

There can be no defence that the substantive statements in the ICoC are a mere reiteration of the provisions already contained in the existing operative space law. The major points stressed in the code are the ban on the weapons and mitigation of space debris in outer space for its long term sustainability. The weapons clause in the code is more cautiously worded than in the Outer Space Treaty[27] (OST) and in addition, the UN Charter has been invoked in self-defence, thus, rendering the statement on prohibition of weapons in outer space completely innocuous, in fact, totally impotent. The comparable Article IV of the OST seems more imperious, with better effect and psychological impact. The OST carries a mandate and not a sheer platitude.[28]

Regarding mitigation of debris, Article IX of the OST is relevant. It may be conceded that it is more objective, oriented and skeletal in approach but it must be appreciated that it was drafted more than three decades earlier and environmental pollution in outer space has since increased many times. Appreciating the gravity of the problem, the Scientific and Technical Sub-Committee of the COPUOS, after long and laborious discussions, has already framed Guidelines on Mitigation of Debris in Outer Space that has been promulgated under the aegis of the UN in 2007. These guidelines, read in collaboration with Article IX of the OST, gain the efficacy of pseudo-compulsory measures binding on all states. In this regard, the code appears sheer rhetoric with no concrete measures suggested therein. Thus, the code is a repetition of existing provisions, albeit in muted and cautious terms, perhaps to gain more adherences at the cost of meaningful substance and earnest intent.

It is, however, advocated that the code appeals to elicit cooperation with transparency in outer space activities like information on launchings and urges for confidence building measures. These are laudable in writing and spirit, and, no doubt, deserve compliance. But parallel provisions are contained in the OST with a mandate for compliance. The relevant Articles X and XI of the OST require states to promote international cooperation in space activities and make it compulsory for states "to inform the Secretary General of the United Nations as well as the public and the international scientific community, to the greatest extent feasible and practicable, of the nature, conduct, location and results of such activities."[29] The OST provision imposes a duty to inform and seems much more meaningful than the exhortation in the code.

In view of the above, there appears nothing new in the ICoC that is not already contained in treaties, various agreements and many UN proclaimed principles and guidelines on different issues. On the contrary, the code merely vouchsafes referential and reverential commitment to the existing legal framework for outer space activities.[30] The reiterations make it an apparently innocuous document with undefined vocabulary on weapons and space disarmament or the nature of debris and methods of disposal or modalities for scavenging of detritus. In fact, it is this utter innocence and lack of actionable narrative that make it so innocuous and hollow. And its urging for responsible national behaviour seems a mere sermon. Despite this simplicity, there are remonstrations to the obvious that have been stated in the ICoC. The irony of the situation is, indeed, lamentable.

Approach is Selective and Not Holistic
The ICoC deals with only a few provisions of the OST relating to "protecting the safety of space operations and the long-term sustainability of outer space operations…enhancing the security of outer space activities by all states and the prevention of an arms race in outer space."[31] This broadly leads to an emphasis on safety of the space assets and sustainability of the space environment by debris mitigation. Another concern highlighted in the code relates to prohibition of weapons in outer space, thus, advocating space disarmament. The code also alludes to certain incidental aspects of operations like registration of space objects, notification of outer space activities, sharing of information, where appropriate, space traffic management and consultative mechanisms on the hazards of operations and conflict-resolution among states.

Therefore, the approach of the code to deal with contemporary space activities is narrow, selective and segmental. The treatment of the subject is not comprehensive. The implication could be to propose separate codes on different subjects. Another guess could be that it is, possibly, dictated by paranoia due to the rising clutter in outer space and the attendant risks to assets and consequent uncertainties in their sustained operations.[32] Of course, the UN Guidelines on Mitigation of Debris in Space have addressed most of the pressing problems and allayed the major fears. Perhaps certain apprehensions remain and, as a result, the focus of the code has not widened beyond this gamut. Undoubtedly, the concern is genuine and equally shared by all yet the apprehension seems misplaced.

For a holistic treatment of the subject, the code must take total stock of contemporary problems ailing or hindering operations in outer space. This would pull the code towards chinks in the law on several issues. Apart from the fact that outer space is becoming congested, causing concerns about safety, there are other contemporary realities that it must take due notice of. For example, new kinds of players have entered the space arena with resounding success like private enterprise and consortium entities such as Space-X, Virgin Galactic, Bigelow, Planetary Resources and many more are waiting in the wings for an entry at an opportune time. Another feasibility that is apparent is mining of celestial resources but its legality is in the shadows and clarity on the issue is lacking. Further, as of today, the ban on Weapons of Mass Destruction (WMDs) is taken for granted under the confidence of the OST but newer military technologies are making their presence felt and the link between Ballistic Missile Defence (BMD) and weaponisation of space is fast evolving. In fact, BMD has moved into the space layer for sensors and interceptors in space could be next. This reality gets demonstrated by China's capability to blind/jam radio frequencies of satellites through lasers in 2006 and then the ASAT test in 2007.[33]

Code is Optional and Non-Binding

The code itself affirms, "Adherence to this Code and to the measures contained in it is voluntary and open to all States."[34] It implies that the signatories or subscribing states have no compulsion to abide by the code and they implicitly reserve an option to defy if any measure in the code is inconvenient to comply with. It is, thus, a *pacta sans servanda* or implying no compulsion to adhere after ascension. It is a travesty of established international law. From this clause emerges the fact that it is a non-binding instrument which makes it weak in enforcement yet attractive as a sheer formality in international relations. From another angle, such a proposal betrays diplomatic diffidence and political timidity to the cause, however, celebrated and fit for espousal. Therefore, a code for peripheral ceremonials with no substantive advantage or compulsion of compliance is hardly worth any commitment.

The intention here is not to hypothesise that soft laws are seldom effective. In fact, empirical studies point to quite the contrary, because human civilisation, from its very dawn, has evolved norms and customs that have held sway with hardly any aberrations in behaviour even in large communities. But over the millennia, attitudes and mindsets have changed and implicit obedience of mores is no longer spontaneous in a beguiled world. Therefore, the underlying assumption of the code that space-faring nations are ethical actors imbued with human virtues who would normally behave within the limits of prescribed discipline appears belied in view of the empirical studies of state practice.

Mere pontification will not make them fall in line and such sermons are rarely taken seriously by political entities. Self-serving national interests take priority; often assume primacy, over wider global causes or the cosmic best for humanity. Hence, such optional codes of non-mandatory provisions overlapping treaty law are confounding and may detract from the credibility of mandates or diffuse their very impact. In practice, such imprecise duplication may encourage misleading interpretations to duck thereunder or even prop up outright mischief. Thus, overlap of a treaty compulsion by a self-resolution would be self-defeating and delusional.

Absence of Definitions

Another serious deficiency of the code is the lack of definition for terms used in the text. For example, words like weapons, outer space, debris, military activities, space traffic management, confidence-building measures *et al* can assume different nuances. In fact, it is an established practice to define important terminology used in international instruments and even municipal statutes. Generally, the purpose of such a provision is to facilitate common understanding and guide its interpretation in the right perspective in furtherance of its avowed objectives.

In an international code, this is all the more necessary because different countries can impute discordant meanings of varying hues due to differences in cultural perceptions and linguistic syntax. This justifies a need for specific definitions related to the code so that all subscribing states accord the same meaning and common connotation to the various words. This is important for an *ad idem* understanding of the code and to avoid innuendoes. This would have helped reinforce confidence-building measures and transparency intended in the code, traits which are not exclusive to space law but inclusive in it. Unfortunately, the OST and other similar instruments also suffer from this typical delinquency.

Code is Not Innovative or Futuristic

The primary deficiency of the code lies in its limited perspective and narrow vision of contemporary problems faced in modern space activities based on advanced scientific applications. Its demerit gets further compounded by duplication of the existing provisions of space law in its contents. Thus, it is hardly innovative or futuristic in its approach and offers little by way of solution to the gaps and grey areas in the existing corpus of space law. The maladies remain gaping for solutions, while the code repeats the established legality.

The code reaffirms a commitment to the existing framework of space law and even lists prevailing conventions and agreements to be more specific. Similarly, principles and guidelines promulgated by the UN have been reiterated. The code also solicits "compliance with and promotion of Treaties, Conventions and other Commitments relating to Outer Space Activities."[35] The subscribing states take a solemn pledge to "reiterate their support to encouraging coordinated efforts in order to promote universal adoption of, and full adherence to, the [space law] instruments."[36] The listing, however, is not complete on the subject and many germane aspects have been ignored. Also, this reveals that the code endorses the *status quo* without being futuristic or proactive in approach. This is hardly commendable.

It is no secret that space law, as it exists today, has many overlaps, gaps as well as grey areas. The gaps like the silence on celestial bodies, including the moon being a common heritage of [hu]mankind and the modalities of sharing of benefits among all states as contained in the Moon Agreement or the connotation of outer space and celestial bodies being the province of mankind as enshrined in the Outer Space Treaty. An example of a grey area is whether private enterprises, on their own, can commercially expropriate natural mineral resources from celestial bodies or helium from the moon or precious metals from near-earth asteroids and the consequent proportional sharing of profits among entitled entities. These and many other such issues seek judicious solutions where legal histrionics hold no answer. The need of the hour is not hollow rhetoric but to evolve space jurisprudence and *jus cogens* of space law.[37]

An illustration of a proactive seizure of a futuristic problem can be to evolve universal rules of space traffic management to be complied with by all space-faring nations. The problem has two dimensions: the first relates to the scientific and technical aspects of the manoeuvrability and disposal of the satellite on completion of the mission or on becoming rogue. The other dimension concerns regulatory field measures for control and supervision of all functions incident to orderly and safe traffic. In other words, regulatory provisions should facilitate safe access into outer space, operations in the space medium and return to the earth free from physical or radio frequency interference. Both need to develop in conjunction for concordance.

This may necessitate setting up of a common space tracking and watch organisation to be funded and maintained by all users, perhaps under the aegis of the UN. All opinions veer to the necessity of a surveillance network for Space Situational Awareness (SSA). The current US Tracking and Information System or independent systems of limited range and utility cannot be relied upon or trusted by all for all times. A common organisation under the UN or a separate World Space Authority with Regional Tracking and Reporting Centres would be more acceptable and cost-effective; and would promote regional cooperation and instill confidence. This enlargement of the functions in a centralised UN organisation would obviate the need for a separate Outer Space Activities Database proposed in the code to collect and disseminate information.[38] Creating a new modality parallel to the UN institutions would be infructuous duplication and, ultimately, may prove to be counter-productive.

A Medley of Opinions

It is now proposed to present a collated array of global views on the code. It reveals a broad spectrum of hues of support and criticism that makes an interesting study.[39] First, since the original Code of Conduct was first presented to the Council of the European Union in 2008, it would be natural to expect the European Union nations to endorse this space initiative, whether strongly or mildly. There has been discussion on the draft and changes proposed but community loyalty expects EU nations to subscribe to the code. Jana Robinson enunciates the rationale. "Europe considers space systems to be strategic assets." And the code has been prompted "... by the troubling display of non-transparency and insensitivity to the space environment..." The logic is, of course, infallible and the cause, worth the pursuit.

Second, though the US has on the sidelines drafted its own code, its support to the EU Code is unequivocal. The US feels that most of the space law was made during the Cold War era "when just two countries conducted almost all the space activity. Today, the context of space activity is changing, as more and more countries and private companies become involved. As a result, various popular orbits are becoming crowded and there is a higher demand for coordination in...the allocation of the radio frequency spectrum, space traffic control, and military systems testing."[40] He adds, "The proposed International Code of Conduct is one means of facilitating such talks. It is not a panacea, but it is a start." Further commenting on the drawbacks of the code, Moltz states, "Given its voluntary nature, the Code will rely largely on political and moral suasion to accomplish its objectives of safer and more predictable space behaviour."[41]

Third, Canada's emphasis focusses on Transparency and Confidence-Building Measures (TCBM) for space security but it reckons that these measures cannot assure "robust security guarantees" and the usefulness of the code lies in generating debate and discussion on a grave concern.

Canada believes that to date, TCBMs have been adopted either as stand-alone actions to reduce mistrust or in combination with other means like complements to treaties. It can be viewed as the foundation or first step to stir momentum for a future legal agreement. Despite this overt support, TCBMs nonetheless, remain mere mental constructs with no concrete basis. A Canadian scholar also acknowledges, "...TCBMs are merely a means of achieving space security but their effectiveness highly depends upon their precise contents, pertinent scope, uniform and effective implementation, objective compliance monitoring and verification, and the presence of efficient dispute settlement mechanisms." The riders are loud and clear to show that the support is a mere formality.[42]

Fourth, Japan's space policy is undergoing a dynamic change because so far its focus was on scientific exploration and technological development; and not operational missions. Hence, it paid little attention to the orbital environment and its constraints. But the situation is changing and a shift in emphasis is discernible. As a result, Japan views the code as a remarkable initiative, suitable for the space-faring community. Its voluntary nature is designed to increase membership but it would be more effective and ideal if it is legally binding. Under the circumstances, "it is not clear how much Japan can contribute to the International Code, but it is certain that Japan is strongly committed."[43] This committal, somehow, appears lukewarm and formalistic.

Another area of influence comprises the Latin American countries and the Caribbean region. Ambassador Ciro Arevalo Yepes representing them, acknowledges that "the space environment is changing rapidly with a growing number of states seeking to develop or extend their space capabilities. A variety of non-state actors are also extending their involvement in space activities." For example, the successful launching of a robotic spacecraft and cargo replenishment missions to the International Space Station by Space-X. "In the new space era that is unfolding [there is]...the need to have a better management of activities in outer space." The scholar feels that the United Nations has been a principal inter-governmental forum on space related issues and should continue to "provide overarching guidance...[to] promote improved coordination and cooperative governance." He further opines that a holistic approach is necessary and, at the same time, comments, "Global space governance should avoid top-down approaches that risk making the proposal inadequate."[44] This opinion rightly treads beyond the current draft of the code to evolve into a holistic document of meaningful content and durable value.

Conclusion

The initiative of the EU is not even a halfway house; it is just a shack when a mansion already exists. May be the mansion needs an extension, a thorough renovation or a minor modification to suit the contemporary demands of technological advances, the adventures of human tolerance and the financial investments that conglomerates are willing to make. Specific maladies need specific remedies and a placebo would be of no help. The world polity should not shy away from, or shirk, this task or the mandate of time as it would amount to sheer timidity and acceptance of defeat even before the encounter. A wise saying puts it that you cannot leave a lasting impression of your footprints if you walk cautiously on tiptoe. The suggestion is loud and clear.

The code is also touted to usher in better international understanding and provide confidence-building measures to reduce the trust-deficit. This premise staggers on the crutches of an assumption that states are ethical actors and would change their attitude to prioritise international interests over national compulsions or alter their perception of national interests. On the other hand, if states were so pliant, their adherence to the mandates of the OST should come unflinchingly and with spontaneity. If this happens, then where is the need for a code? Such a change of heart cannot be accomplished by rhetoric and appeals. In a nutshell, despite the crescendo of approval for the code, its advantage appears illusory and suspect.

In conclusion, the code appears best suited as a modality to induce cooperation and consultation among the European Union nations, as the idea was originally mooted, rather than be sponsored as an International Code of Conduct, for reasons of its doubtful value and alleged superfluity. Further, the option of voluntary adherence offered as bait in the code detracts from its true merit. Nevertheless, having garnered a groundswell of support and marshalled endorsements, it will not be wise to fritter away this opportunity. The problems must be acknowledged and viewed in the right perspective for amelioration. Perhaps, it could be more efficacious if, to begin with, the code operates within the EU and is made binding and obligatory within the EU. It should, thus, work as a catalyst for change. In such an eventuality, it could demonstrate the constructive interplay of the code and the European Community could present an example of cooperation among nations and foster betterment of humanity. It should then be emulated by the world polity. It is wisely said, example is better than precept.

Notes

1. Manpreet Sethi, "Regulating Human Activities in Outer Space—Is Code of Conduct the Answer?" *Asian Defence Review 2012* (CAPS), (New Delhi: KW Publishers Pvt Ltd, 2013), p. 22.
2. Generally, refer to Jana Robinson, "Europe's Space Diplomacy Initiative: The International Code of Conduct" in Ajay Lele, ed., *Decoding the International Code of Conduct for Outer Space Activities* (Pentagon Security International, 2012), pp. 27-29.
3. Sourced from http://www.consilium.europa.eu/uedocs/cmsUpload/st14455.en10.pdf.
4. Lele, ed., n. 2, p. xvii.
5. EU Press Note issued on the subject from Brussels on June 6, 2012.
6. Refer Ibid.
7. Ibid., Paragraph 2. Words in parentheses added for clarity in understanding.
8. Text of International Code of Conduct for Outer Space Activities (2010) presented by the Council of the European Union. Introductory paragraph 2. Accessible at http://www.consilium.europa.eu/uedoc/cmsUpload/st14455.en10.pdf
9. International Code of Conduct for Outer Space Activities (2010), Preamble. Words in parentheses added/changed for clarity.
10. International Code of Conduct for Outer Space Activities (2010), Preamble.
11. International Code of Conduct for Outer Space Activities (2010), Section 1, paragraph 1.
12. Nancy Gallagher, "International Cooperation and Space Governance Strategy" in Eligar Sadeh, ed., *Space Strategy in the 21st Century, Theory and Practice,* (Routledge, 2013), p.61.
13. It would be mischievous to dub them as the Communist Bloc.
14. Draft of PPWT dated February 12, 2008. Gallagher, n. 12, fn. 26 on p.74.

15. Gallagher, n. 12, pp. 61-62.
16. For more details on the subject, refer Rajeswari Pillai Rajagopalan, "Debate on Space Code of Conduct: An Indian Perspective", *ORF Occasional Paper # 26,* October 2011, Observer Resaerch Foundation, New Delhi.
17. For different shades of opinion, refer Lele, n. 2.
18. Treaty on Principles Governing the Activities of States in the Exploration and Use of Outer Space, including the Moon and other Celestial Bodies, 1967. It entered into force on October 10, 1967.
19. Refer Preamble to the Model Code of Conduct (MCoC).
20. The Outer Space Treaty, Article III.
21. Refer Michael Listner, "An Update on the Proposed European Code of Conduct", *The Space Review,* August, 08, 2011.
22. Lele, ed., n. 2, p. 6.
23. Ibid., p. xix. Words in parentheses are added.
24. ICoC (2010), Preamble.
25. For related treatment of the subject, refer "The Space Doctrine : Agenda for the UN," in G. S. Sachdeva, *Outer Space—Security and Legal Challenges* (New Delhi: Knowledge World Publishers, 2010), pp. 57-88.
26 (2010), Section-III Cooperation Mechanisms, paragraph 6.
27. Treaty on Principles Governing the Activities of States in the Exploration and Use of Outer Space including the Moon and the other Celestial Bodies, 1969.
28. For more detailed analysis, refer chapter on Weaponisation of Outer Space and National Security in Sachdeva, n. 25, pp. 161-186.
29. Treaty on Principles Governing the Activities of States in the Exploration and Use of Outer Space including the Moon and the other Celestial Bodies, 1969. Article XI.
30. ICoC (2010), Section I, Purpose, Scope and Core Principles, paragraph 3.1.
31. Text of ICoC (2010), Section II, General Measures, paragraphs 4.4 and 4.5.
32. Refer ICoC (2010), Section III, Cooperative Mechanisms, paragraph 6.1.
33. For a detailed treatment, refer Sethi, n. 1.
34. Refer ICoC (2010), Section III, Cooperative Mechanisms, paragraph 1. 4.
35. Refer ICoC (2010), paragraph 3.
36. Ibid., paragraph 3.2.
37. G. S. Sachdeva, "Jus Cogens of Space Law," *Asian Journal of Air and Space Law,* Vol II, No 2, July-December, 2012.
38. ICoC (2010), refer paragraph 12.
39. Most of the views have been culled out and collated from Lele, eds., n. 2. Individual citations avoided in this section.
40. James Clay Moltz, "The Code of Conduct: A Useful First Step" in Lele, ed., n. 2, pp. 43-45.

41. Ibid.
42. Views abstracted from Ram Jakhu, "Transparency and Confidence-Building Measures for Space Security", in Lele, ed., n. 2, pp. 35-46.
43. Kazuto Suzuki, "Japan, Space Security and Code of Conduct," in Lele, ed., n. 2, pp. 94-96.
44. Ciro Arevalo Yepes, "CoC: Need for a Holistic Approach", in Lele, ed., n. 2, pp. 122-124.

Annexure A

United Nations Treaties and Principles on Outer Space (United Nations New York, 2002)

A. Treaty on Principles Governing the Activities of States in the Exploration and Use of Outer Space, including the Moon and Other Celestial Bodies

The States Parties to this Treaty,

Inspired by the great prospects opening up before mankind as a result of man's entry into outer space,

Recognizing the common interest of all mankind in the progress of the exploration and use of outer space for peaceful purposes,

Believing that the exploration and use of outer space should be carried on for the benefit of all peoples irrespective of the degree of their economic or scientific development,

Desiring to contribute to broad international cooperation in the scientific as well as the legal aspects of the exploration and use of outer space for peaceful purposes,

Believing that such cooperation will contribute to the development of mutual understanding and to the strengthening of friendly relations between States and peoples,

Recalling resolution 1962 (XVIII), entitled "Declaration of Legal Principles Governing the Activities of States in the Exploration and Use of Outer Space", which was adopted unanimously by the United Nations General Assembly on 13 December 1963,

Recalling resolution 1884 (XVIII), calling upon States to refrain from placing in orbit around the Earth any objects carrying nuclear weapons or any other kinds of weapons of mass destruction or from installing such weapons on celestial bodies, which was adopted unanimously by the United Nations General Assembly on 17 October 1963,

Taking account of United Nations General Assembly resolution 110 (II) of 3 November 1947, which condemned propaganda designed or likely to provoke or encourage any threat to the peace, breach of the peace or act of aggression, and considering that the aforementioned resolution is applicable to outer space,

Convinced that a Treaty on Principles Governing the Activities of States in the Exploration and Use of Outer Space, including the Moon and Other Celestial Bodies, will further the purposes and principles of the Charter of the United Nations,

Have agreed on the following:

Article I

The exploration and use of outer space, including the Moon and other celestial bodies, shall be carried out for the benefit and in the interests of all countries, irrespective of their degree of economic or scientific development, and shall be the province of all mankind.

Outer space, including the Moon and other celestial bodies, shall be free for exploration and use by all States without discrimination of any kind, on a basis of equality and in accordance with international law, and there shall be free access to all areas of celestial bodies.

There shall be freedom of scientific investigation in outer space, including the

Moon and other celestial bodies, and States shall facilitate and encourage international cooperation in such investigation.

Article II

Outer space, including the Moon and other celestial bodies, is not subject to national appropriation by claim of sovereignty, by means of use or occupation, or by any other means.

Article III

States Parties to the Treaty shall carry on activities in the exploration and use of outer space, including the Moon and other celestial bodies, in accordance with international law, including the Charter of the United Nations, in the interest of maintaining international peace and security and promoting international cooperation and understanding.

Article IV

States Parties to the Treaty undertake not to place in orbit around the Earth any objects carrying nuclear weapons or any other kinds of weapons of mass destruction, install such weapons on celestial bodies, or station such weapons in outer space in any other manner.

The Moon and other celestial bodies shall be used by all States Parties to the Treaty exclusively for peaceful purposes. The establishment of military bases, installations and fortifications, the testing of any type of weapons and the conduct of military manoeuvres on celestial bodies shall be forbidden. The use of military personnel for scientific research or for any other peaceful purposes shall not be prohibited. The use of any equipment or facility necessary for peaceful exploration of the Moon and other celestial bodies shall also not be prohibited.

Article V

States Parties to the Treaty shall regard astronauts as envoys of mankind in outer space and shall render to them all possible assistance in the event of accident, distress, or emergency landing on the territory of another State Party or on the high seas. When astronauts make such a landing, they shall be safely and promptly returned to the State of registry of their space vehicle.

In carrying on activities in outer space and on celestial bodies, the astronauts of one State Party shall render all possible assistance to the astronauts of other States Parties.

States Parties to the Treaty shall immediately inform the other States Parties to the Treaty or the Secretary-General of the United Nations of any phenomena they discover in outer space, including the Moon and other celestial bodies, which could constitute a danger to the life or health of astronauts.

Article VI
States Parties to the Treaty shall bear international responsibility for national activities in outer space, including the Moon and other celestial bodies, whether such activities are carried on by governmental agencies or by non-governmental entities, and for assuring that national activities are carried out in conformity with the provisions set forth in the present Treaty. The activities of non-governmental entities in outer space, including the Moon and other celestial bodies, shall require authorization and continuing supervision by the appropriate State Party to the Treaty. When activities are carried on in outer space, including the Moon and other celestial bodies, by an international organization, responsibility for compliance with this Treaty shall be borne both by the international organization and by the States Parties to the Treaty participating in such organization.

Article VII
Each State Party to the Treaty that launches or procures the launching of an object into outer space, including the Moon and other celestial bodies, and each State Party from whose territory or facility an object is launched, is internationally liable for damage to another State Party to the Treaty or to its natural or juridical persons by such object or its component parts on the Earth, in air space or in outer space, including the Moon and other celestial bodies.

Article VIII
A State Party to the Treaty on whose registry an object launched into outer space is carried shall retain jurisdiction and control over such object, and over any personnel thereof, while in outer space or on a celestial body. Ownership of objects launched into outer space, including objects landed or constructed on a celestial body, and of their component parts, is not affected by their presence in outer space or on a celestial body or by their return to the Earth. Such objects or component parts found beyond the limits of the State Party to the Treaty on whose registry they are carried shall be returned to that State Party, which shall, upon request, furnish identifying data prior to their return.

Article IX

In the exploration and use of outer space, including the Moon and other celestial bodies, States Parties to the Treaty shall be guided by the principle of cooperation and mutual assistance and shall conduct all their activities in outer space, including the Moon and other celestial bodies, with due regard to the corresponding interests of all other States Parties to the Treaty. States Parties to the Treaty shall pursue studies of outer space, including the Moon and other celestial bodies, and conduct exploration of them so as to avoid their harmful contamination and also adverse changes in the environment of the Earth resulting from the introduction of extraterrestrial matter and, where necessary, shall adopt appropriate measures for this purpose. If a State Party to the Treaty has reason to believe that an activity or experiment planned by it or its nationals in outer space, including the Moon and other celestial bodies, would cause potentially harmful interference with activities of other States Parties in the peaceful exploration and use of outer space, including the Moon and other celestial bodies, it shall undertake appropriate international consultations before proceeding with any such activity or experiment. A State Party to the Treaty which has reason to believe that an activity or experiment planned by another State Party in outer space, including the Moon and other celestial bodies, would cause potentially harmful interference with activities in the peaceful exploration and use of outer space, including the Moon and other celestial bodies, may request consultation concerning the activity or experiment.

Article X

In order to promote international cooperation in the exploration and use of outer space, including the Moon and other celestial bodies, in conformity with the purposes of this Treaty, the States Parties to the Treaty shall consider on a basis of equality any requests by other States Parties to the Treaty to be afforded an opportunity to observe the flight of space objects launched by those States.

The nature of such an opportunity for observation and the conditions under which it could be afforded shall be determined by agreement between the States concerned.

Article XI

In order to promote international cooperation in the peaceful exploration and use of outer space, States Parties to the Treaty conducting activities in outer space, including the Moon and other celestial bodies, agree to inform the Secretary-General of the United Nations as well as the public and the international scientific community, to the greatest extent feasible and practicable, of the nature, conduct, locations and results of such activities. On receiving the said information, the Secretary-General of the United Nations should be prepared to disseminate it immediately and effectively.

Article XII

All stations, installations, equipment and space vehicles on the Moon and other celestial bodies shall be open to representatives of other States Parties to the Treaty on a basis of reciprocity. Such representatives shall give reasonable advance notice of a projected visit, in order that appropriate consultations may be held and that maximum precautions may be taken to assure safety and to avoid interference with normal operations in the facility to be visited.

Article XIII

The provisions of this Treaty shall apply to the activities of States Parties to the Treaty in the exploration and use of outer space, including the Moon and other celestial bodies, whether such activities are carried on by a single State Party to the Treaty or jointly with other States, including cases where they are carried on within the framework of international intergovernmental organizations.

Any practical questions arising in connection with activities carried on by international intergovernmental organizations in the exploration and use of outer space, including the Moon and other celestial bodies, shall be resolved by the States Parties to the Treaty either with the appropriate international organization or with one or more States members of that international organization, which are Parties to this Treaty.

Article XIV

This Treaty shall be open to all States for signature. Any State which does not sign this Treaty before its entry into force in accordance with paragraph 3 of this article may accede to it at any time.

This Treaty shall be subject to ratification by signatory States. Instruments of ratification and instruments of accession shall be deposited with the Governments of the Union of Soviet Socialist Republics, the United Kingdom of Great Britain and Northern Ireland and the United States of America, which are hereby designated the Depositary Governments.

This Treaty shall enter into force upon the deposit of instruments of ratification by five Governments including the Governments designated as Depositary Governments under this Treaty.

For States whose instruments of ratification or accession are deposited subsequent to the entry into force of this Treaty, it shall enter into force on the date of the deposit of their instruments of ratification or accession.

The Depositary Governments shall promptly inform all signatory and acceding States of the date of each signature, the date of deposit of each instrument of ratification of and accession to this Treaty, the date of its entry into force and other notices.

This Treaty shall be registered by the Depositary Governments pursuant to Article 102 of the Charter of the United Nations.

Article XV

Any State Party to the Treaty may propose amendments to this Treaty. Amendments shall enter into force for each State Party to the Treaty accepting the amendments upon their acceptance by a majority of the States Parties to the Treaty and thereafter for each remaining State Party to the Treaty on the date of acceptance by it.

Article XVI

Any State Party to the Treaty may give notice of its withdrawal from the Treaty one year after its entry into force by written notification to the Depositary Governments. Such withdrawal shall take effect one year from the date of receipt of this notification.

Article XVII

This Treaty, of which the Chinese, English, French, Russian and Spanish texts are equally authentic, shall be deposited in the archives of the Depositary Governments. Duly certified copies of this Treaty shall be transmitted by the Depositary Governments to the Governments of the signatory and acceding States.

IN WITNESS WHEREOF the undersigned, duly authorized, have signed this Treaty.

DONE in triplicate, at the cities of London, Moscow and Washington, D.C., the twenty-seventh day of January, one thousand nine hundred and sixty-seven.

Annexure B

Agreement Governing the Activities of States on the Moon and Other Celestial Bodies (1979)

ENTERED INTO FORCE: 11 July 1984

The States Parties to this Agreement,

Noting the achievements of States in the exploration and use of the moon and other celestial bodies,

Recognizing that the moon, as a natural satellite of the earth, has an important role to play in the exploration of outer space,

Determined to promote on the basis of equality the further development of co-operation among States in the exploration and use of the moon and other celestial bodies,

Desiring to prevent the moon from becoming an area of international conflict,

Bearing in mind the benefits which may be derived from the exploitation of the natural resources of the moon and other celestial bodies,

Recalling the Treaty on Principles Governing the Activities of States in the Exploration and Use of Outer Space, including the Moon and Other Celestial Bodies, the Agreement on the Rescue of Astronauts, the Return of Astronauts and the Return of Objects Launched into Outer Space, the Convention on International Liability for Damage Caused by Space Objects, and the Convention on Registration of Objects Launched into Outer Space,

Taking into account the need to define and develop the provisions of these international instruments in relation to the moon and other celestial bodies, having regard to further progress in the exploration and use of outer space,

Have agreed on the following:

Article 1

1. The provisions of this Agreement relating to the moon shall also apply to other celestial bodies within the solar system, other than the earth, except in so far as specific legal norms enter into force with respect to any of these celestial bodies.
2. For the purposes of this Agreement reference to the moon shall include orbits around or other trajectories to or around it.
3. This Agreement does not apply to extraterrestrial materials which reach the surface of the earth by natural means.

Article 2

All activities on the moon, including its exploration and use, shall be carried out in accordance with international law, in particular the Charter of the United Nations, and taking into account the Declaration on Principles of International Law concerning Friendly Relations and Co-operation Among States in accordance with the Charter of the United Nations, adopted by the General Assembly on 24 October 1970, in the interests of maintaining international peace and security and promoting international co-operation and mutual understanding, and with due regard to the corresponding interests of all other States Parties.

Article 3

1. The moon shall be used by all States Parties exclusively for peaceful purposes.
2. Any threat or use of force or any other hostile act or threat of hostile act on the moon is prohibited. It is likewise prohibited to use the moon in order to commit any such act or to engage in any such threat in relation to the earth, the moon, spacecraft, the personnel of spacecraft or man-made space objects.

3. States Parties shall not place in orbit around or other trajectory to or around the moon objects carrying nuclear weapons or any other kinds of weapons of mass destruction or place or use such weapons on or in the moon.
4. The establishment of military bases, installations and fortifications, the testing of any type of weapons and the conduct of military manoeuvres on the moon shall be forbidden. The use of military personnel for scientific research or for any other peaceful purposes shall not be prohibited. The use of any equipment or facility necessary for peaceful exploration and use of the moon shall also not be prohibited.

Article 4

1. The exploration and use of the moon shall be the province of all mankind and shall be carried out for the benefit and in the interests of all countries, irrespective of their degree of economic or scientific development. Due regard shall be paid to the interests of present and future generations as well as to the need to promote higher standards of living and conditions of economic and social progress and development in accordance with the Charter of the United Nations.
2. States Parties shall be guided by the principle of co-operation and mutual assistance in all their activities concerning the exploration and use of the moon. International co-operation in pursuance of this Agreement should be as wide as possible and may take place on a multilateral basis, on a bilateral basis or through international intergovernmental organizations.

Article 5

1. States Parties shall inform the Secretary-General of the United Nations as well as the public and the international scientific community, to the greatest extent feasible and practicable, of their activities concerned with the exploration and use of the moon. Information on the time, purposes, locations, orbital parameters and duration shall be given in respect of each mission to the moon as soon as possible after launching, while information on the results of each mission, including scientific results, shall be furnished upon completion of the mission. In the case of a mission lasting more than thirty days, information on

conduct of the mission, including any scientific results, shall be given periodically at thirty days' intervals. For missions lasting more than six months, only significant additions to such information need be reported thereafter.
2. If a State Party becomes aware that another State Party plans to operate simultaneously in the same area of or in the same orbit around or trajectory to or around the moon, it shall promptly inform the other State of the timing of and plans for its own operations.
3. In carrying out activities under this Agreement, States Parties shall promptly inform the Secretary-General, as well as the public and the international scientific community, of any phenomena they discover in outer space, including the moon, which could endanger human life or health, as well as of any indication of organic life.

Article 6
1. There shall be freedom of scientific investigation on the moon by all States Parties without discrimination of any kind, on the basis of equality and in accordance with international law.
2. In carrying out scientific investigations and in furtherance of the provisions of this Agreement, the States Parties shall have the right to collect on and remove from the moon samples of its mineral and other substances. Such samples shall remain at the disposal of those States Parties which caused them to be collected and may be used by them for scientific purposes. States Parties shall have regard to the desirability of making a portion of such samples available to other interested States Parties and the international scientific community for scientific investigation. States Parties may in the course of scientific investigations also use mineral and other substances of the moon in quantities appropriate for the support of their missions.
3. States Parties agree on the desirability of exchanging scientific and other personnel on expeditions to or installations on the moon to the greatest extent feasible and practicable.

Article 7
1. In exploring and using the moon, States Parties shall take measures to prevent the disruption of the existing balance of its environment whether

by introducing adverse changes in that environment, by its harmful contamination through the introduction of extra-environmental matter or otherwise. States Parties shall also take measures to avoid harmfully affecting the environment of the earth through the introduction of extraterrestrial matter or otherwise.

2. States Parties shall inform the Secretary-General of the United Nations of the measures being adopted by them in accordance with paragraph 1 of this article and shall also, to the maximum extent feasible, notify him in advance of all placements by them of radio-active materials on the moon and of the purposes of such placements.

3. States Parties shall report to other States Parties and to the Secretary-General concerning areas of the moon having special scientific interest in order that, without prejudice to the rights of other States Parties, consideration may be given to the designation of such areas as international scientific preserves for which special protective arrangements are to be agreed upon in consultation with the competent bodies of the United Nations.

Article 8

1. States Parties may pursue their activities in the exploration and use of the moon anywhere on or below its surface, subject to the provisions of this Agreement.

2. For these purposes States Parties may, in particular:
   (a) Land their space objects on the moon and launch them from the moon;
   (b) Place their personnel, space vehicles, equipment, facilities, stations and installations anywhere on or below the surface of the moon.

Personnel, space vehicles, equipment, facilities, stations and installations may move or be moved freely over or below the surface of the moon.

3. Activities of States Parties in accordance with paragraphs 1 and 2 of this article shall not interfere with the activities of other States Parties on the moon. Where such interference may occur, the States Parties concerned shall undertake consultations in accordance with article 15, paragraphs 2 and 3 of this Agreement.

Article 9

1. States Parties may establish manned and unmanned stations on the moon. A State Party establishing a station shall use only that area which is required for the needs of the station and shall immediately inform the Secretary-General of the United Nations of the location and purposes of that station. Subsequently, at annual intervals that State shall likewise inform the Secretary-General whether the station continues in use and whether its purposes have changed.
2. Stations shall be installed in such a manner that they do not impede the free access to all areas of the moon by personnel, vehicles and equipment of other States Parties conducting activities on the moon in accordance with the provisions of this Agreement or of article I of the Treaty on Principles Governing the Activities of States in the Exploration and Use of Outer Space, including the Moon and Other Celestial Bodies.

Article 10

1. States Parties shall adopt all practicable measures to safeguard the life and health of persons on the moon. For this purpose they shall regard any person on the moon as an astronaut within the meaning of article V of the Treaty on Principles Governing the Activities of States in the Exploration and Use of Outer Space, including the Moon and Other Celestial Bodies and as part of the personnel of a spacecraft within the meaning of the Agreement on the Rescue of Astronauts, the Return of Astronauts and the Return of Objects Launched into Outer Space.
2. States Parties shall offer shelter in their stations, installations, vehicles and other facilities to persons in distress on the moon.

Article 11

1. The moon and its natural resources are the common heritage of mankind, which finds its expression in the provisions of this Agreement and in particular in paragraph 5 or this article.
2. The moon is not subject to national appropriation by any claim of sovereignty, by means of use or occupation, or by any other means.

3. Neither the surface nor the subsurface of the moon, nor any part thereof or natural resources in place, shall become property of any State, international intergovernmental or non-governmental organization, national organization or non-governmental entity or of any natural person. The placement of personnel, space vehicles, equipment, facilities, stations and installations on or below the surface of the moon, including structures connected with its surface or subsurface, shall not create a right of ownership over the surface or the subsurface of the moon or any areas thereof. The foregoing provisions are without prejudice to the international regime referred to in paragraph 5 of this article.
4. States Parties have the right to exploration and use of the moon without discrimination of any kind, on a basis of equality and in accordance with international law and the terms of this Agreement.
5. States Parties to this Agreement hereby undertake to establish an international regime, including appropriate procedures, to govern the exploitation of the natural resources of the moon as such exploitation is about to become feasible. This provision shall be implemented in accordance with article 18 of this Agreement.
6. In order to facilitate the establishment of the international regime referred to in paragraph 5 of this article, States Parties shall inform the Secretary-General of the United Nations as well as the public and the international scientific community, to the greatest extent feasible and practicable, of any natural resources they may discover on the moon.
7. The main purposes of the international regime to be established shall include:
   (a) The orderly and safe development of the natural resources of the moon;
   (b) The rational management of those resources;
   (c) The expansion of opportunities in the use of those resources;
   (d) An equitable sharing by all States Parties in the benefits derived from those resources, whereby the interests and needs of the developing countries, as well as the efforts of those countries which have contributed either directly or indirectly to the exploration of the moon, shall be given special consideration.

8. All the activities with respect to the natural resources of the moon shall be carried out in a manner compatible with the purposes specified in paragraph 7 of this article and the provisions of article 6, paragraph 2, of this Agreement.

Article 12

1. States Parties shall retain jurisdiction and control over their personnel, vehicles, equipment, facilities, stations and installations on the moon. The ownership of space vehicles, equipment, facilities, stations and installations shall not be affected by their presence on the moon.
2. Vehicles, installations and equipment or their component parts found in places other than their intended location shall be dealt with in accordance with article 5 of the Agreement on Rescue of Astronauts, the Return of Astronauts and the Return of Objects Launched into Outer Space.
3. In the event of an emergency involving a threat to human life, States Parties may use the equipment, vehicles, installations, facilities or supplies of other States Parties on the moon. Prompt notification of such use shall be made to the Secretary-General of the United Nations or the State Party concerned.

Article 13

A State Party which learns of the crash landing, forced landing or other unintended landing on the moon of a space object, or its component parts, that were not launched by it, shall promptly inform the launching State Party and the Secretary-General of the United Nations.

Article 14

1. States Parties to this Agreement shall bear international responsibility for national activities on the moon, whether such activities are carried on by governmental agencies or by non-governmental entities, and for assuring that national activities are carried out in conformity with the provisions set forth in this Agreement. States Parties shall ensure that non-governmental entities under their jurisdiction shall engage

in activities on the moon only under the authority and continuing supervision of the appropriate State Party.
2. States Parties recognize that detailed arrangements concerning liability for damage caused on the moon, in addition to the provisions of the Treaty on Principles Governing the Activities of States in the Exploration and Use of Outer Space, including the Moon and Other Celestial Bodies and the Convention on International Liability for Damage Caused by Space Objects, may become necessary as a result of more extensive activities on the moon. Any such arrangements shall be elaborated in accordance with the procedure provided for in article 18 of this Agreement.

Article 15
1. Each State Party may assure itself that the activities of other States Parties in the exploration and use of the moon are compatible with the provisions of this Agreement. To this end, all space vehicles, equipment, facilities, stations and installations on the moon shall be open to other States Parties. Such States Parties shall give reasonable advance notice of a projected visit, in order that appropriate consultations may be held and that maximum precautions may be taken to assure safety and to avoid interference with normal operations in the facility to be visited. In pursuance of this article, any State Party may act on its own behalf or with the full or partial assistance of any other State Party or through appropriate international procedures within the framework of the United Nations and in accordance with the Charter.
2. A State Party which has reason to believe that another State Party is not fulfilling the obligations incumbent upon it pursuant to this Agreement or that another State Party is interfering with the rights which the former State has under this Agreement may request consultations with that State Party. A State Party receiving such a request shall enter into such consultations without delay. Any other State Party which requests to do so shall be entitled to take part in the consultations. Each State Party participating in such consultations shall seek a mutually acceptable resolution of any controversy and shall bear in mind the rights and interests of all States Parties. The Secretary-General of the United

Nations shall be informed of the results of the consultations and shall transmit the information received to all States Parties concerned.
3. If the consultations do not lead to a mutually acceptable settlement which has due regard for the rights and interests of all States Parties, the parties concerned shall take all measures to settle the dispute by other peaceful means of their choice appropriate to the circumstances and the nature of the dispute. If difficulties arise in connexion with the opening of consultations or if consultations do not lead to a mutually acceptable settlement, any State Party may seek the assistance of the Secretary-General, without seeking the consent of any other State Party concerned, in order to resolve the controversy. A State Party which does not maintain diplomatic relations with another State Party concerned shall participate in such consultations, at its choice, either itself or through another State Party or the Secretary-General as intermediary.

Article 16
With the exception of articles 17 to 21, references in this Agreement to States shall be deemed to apply to any international intergovernmental organization which conducts space activities if the organization declares its acceptance of the rights and obligations provided for in this Agreement and if a majority of the States members of the organization are States Parties to this Agreement and to the Treaty on Principles Governing the Activities of States in the Exploration and Use of Outer Space, including the Moon and Other Celestial Bodies. States members of any such organization which are States Parties to this Agreement shall take all appropriate steps to ensure that the organization makes a declaration in accordance with the foregoing.

Article 17
Any State Party to this Agreement may propose amendments to the Agreement. Amendments shall enter into force for each State Party to the Agreement accepting the amendments upon their acceptance by a majority of the States Parties to the Agreement and thereafter for each remaining State Party to the Agreement on the date of acceptance by it.

Article 18

Ten years after the entry into force of this Agreement, the question of the review of the Agreement shall be included in the provisional agenda of the General Assembly of the United Nations in order to consider, in the light of past application of the Agreement, whether it requires revision. However, at any time after the Agreement has been in force for five years, the Secretary-General of the United Nations, as depository, shall, at the request of one third of the States Parties to the Agreement and with the concurrence of the majority of the States Parties, convene a conference of the States Parties to review this Agreement. A review conference shall also consider the question of the implementation of the provisions of article 11, paragraph 5, on the basis of the principle referred to in paragraph 1 of that article and taking into account in particular any relevant technological developments.

Article 19

1. This Agreement shall be open for signature by all States at United Nations Headquarters in New York.
2. This Agreement shall be subject to ratification by signatory States. Any State which does not sign this Agreement before its entry into force in accordance with paragraph 3 of this article may accede to it at any time. Instruments of ratification or accession shall be deposited with the Secretary-General of the United Nations.
3. This Agreement shall enter into force on the thirtieth day following the date of deposit of the fifth instrument of ratification.
4. For each State depositing its instrument of ratification or accession after the entry into force of this Agreement, it shall enter into force on the thirtieth day following the date of deposit of any such instrument.
5. The Secretary-General shall promptly inform all signatory and acceding States of the date of each signature, the date of deposit of each instrument of ratification or accession to this Agreement, the date of its entry into force and other notices.

Article 20

Any State Party to this Agreement may give notice of its withdrawal from the Agreement one year after its entry into force by written notification to the Secretary-General of the United Nations. Such withdrawal shall take effect one year from the date of receipt of this notification.

Article 21

The original of this Agreement, of which the Arabic, Chinese, English, French, Russian and Spanish texts are equally authentic, shall be deposited with the Secretary-General of the United Nations, who shall send certified copies thereof to all signatory and acceding States.

In witness whereof the undersigned, being duly authorized thereto by their respective Governments, have signed this Agreement, opened for signature at New York on December 18, 1979.

Index

Afghanistan, 80, 104, 105, 106, 116
Algeria, 86, 113
Andaman and Nicobar Islands, 75
Antarctica, 49, 191, 195
Apollo, 143, 159, 161, 187
Argentina, 86
Armstrong, Neil, 32
Aryabhatta, 65, 84, 91
Asia, 86, 88, 104, 105, 106, 116, 120
Asia-Pacific Regional Space Agency Forum (APRSAF), 86
Association of Southeast Asian Nations (ASEAN), 86, 104
Asteroids, 157, 158, 159, 160, 161, 162, 163, 164, 169, 170, 174, 175, 180, 181, 182, 183, 184, 185, 186, 187, 189, 190, 191, 192, 197, 200, 226
Astronauts, 21, 31, 32, 34, 43, 44, 45, 46, 47, 52, 58, 130, 147, 216
Astronauts as Envoys of Mankind in Outer Space: Resolution of Dilemma, 31-59
   clarifications in abundanti cautela, 46-51
      passengers are not astronauts, 46
      personnel on unlawful missions not to be envoys, 46-47
   conclusion, 57-59
   definition of astronaut, 34-43
      Chinese nomenclature, 37-38
      Indian term, 38-40
      necessity, 34-35
      robonauts, 40-42
      Russian terminology, 35-37
      US definition, 40
      varying terminology in space law, 42-43
   dilemma and its resolution, 53-57
   introduction, 32-34
   legal appraisal, 51-53
   proposed definition, 43-45
      amplification, 44-45
      criteria, 43-44
Australia, 84, 85, 136, 214

Ballistic Missile Defence (BMD), 88, 90, 114, 223
Bangladesh, 104, 105, 106, 115, 117

Beijing, 89
Bhutan, 104, 106, 116, 117
Bolivia, 117
Bombay, 65
Brazil, 84, 85, 103
Bulgaria, 145

Calcutta, 66
Canada, 85, 145, 227, 228
Celestial Bodies, 2, 7, 9, 15, 16, 17, 18, 19, 20, 21, 25, 26, 34, 54, 56, 57, 65, 80, 125, 127, 129, 30, 131, 32, 137, 147, 155, 156, 157, 159, 161, 165, 167, 169, 170, 171, 172, 173, 174, 175, 176, 178, 179, 180, 181, 182, 183, 184, 186, 188, 189, 190, 191, 192, 197, 198, 199, 200, 206, 217, 226
Chandrayaan-I, 63, 77
Chandrayaan-II, 63, 72, 77, 85, 91
Chicago Convention, 58
Chile, 86
China, 63, 75, 78, 85, 87, 88, 89, 90, 92, 93, 95, 103, 104, 113, 114, 116, 117, 118, 136, 146, 149, 208, 212, 220
Christ, 36, 37
Christian, 36, 37, 124
Cold War, 52, 56, 57, 115, 141, 142, 143, 213, 227
Committee on Peaceful Uses of Outer Space (COPUOS), 1, 51, 66, 127, 138, 146, 147, 207, 215, 216, 221
Common Heritage of Mankind (CHM), 181, 183, 192, 193, 194, 195, 196
Communications, Computers, Intelligence, Surveillance, Reconnaissance (C4ISR), 89, 114
Conference on Disarmament (CD), 80, 212
Cuba, 146
Czechoslovakia, 146

Department of Space (DOS), 64, 69
European Union (EU) (Europe), 23, 87, 104, 113, 134, 149, 206, 208, 209, 227, 229

Facebook, 101, 108
France, 84, 85, 208

GAGAN (GPS Aided Geo-Augmented
Navigation), 63, 71, 75, 87
Gagarin, Yuri, 32, 36, 37
Geo- Synchronous Launch Vehicle (GSLV),
71, 75
Gmail, 101
Germany, 81, 85
Global Positioning System (GPS), 63, 71, 72, 75

Hope, Bill, 167
Hotmail, 101
Hungary, 86, 145
India, 38, 63-95, 101-120, 135, 136, 145, 146,
149, 165, 198, 209, 214, 215
   as vendor of space utilities to developing
   countries, an example in cooperation, 101-120
      benefits from space utilities, 108-110
      competitive alternative, 113-116
      conclusion, 119, 120
      introduction, 101-103
      occasional insensitivity of india towards
      developing countries, 116-119
         case of nepal, 116-117
         lessons, 118-119
         other such cases, 117-118
      organisations of developing countries,
      103-107
         asian organisations, 103-104
         development needs of developing
         countries, 105-107
         other regional organisations, 105
   space leadership of india, 110-113
   few pointers, space policy of, 63-95
Indian National Satellite (INSAT), 69, 75, 76, 77
Indian Remote Sensing (IRS), 69, 84, 86, 91, 112
Indian Space Research Organisation (ISRO),
38, 64, 66, 67, 71, 72, 73, 74, 78, 82, 83, 85,
86, 107, 112, 115, 116, 117, 118, 119, 145
International Civil Aviation Organisation
(ICAO), 198
International Code of Conduct for Activities
in Outer Space: An Exercise in Futility,
205-230
   absence of definitions, 224-225
   approach is selective and not holistic, 222-223
   code is not innovative or futuristic, 225-227
   code is optional and non-binding, 223-224
   conclusion, 229-230
   critical appraisal, 220
   existing corpus of space law, 215-218
      allied treaties regulating outer space
      activities, 216-217
      principles, guidelines and declaration,
      217-218
      united nations charter, 217
   favourable appraisal, 218-220
   international code of conduct, 207-211
      evolution, 207-209
      motivations for code, 210-211
   introduction, 205-207
   medley of opinions, 227-229
   other versions, 210-215
      asian approach, 213-214
      chinese variant, 212-213
      indian stand, 214-215
      united states, 210-212
   repetition of existing provisions, 221-222
International Cooperation As Core Concept of
Space Law: For Diplomacy and Confidence,
123-151
   code of conduct for outer space activities,
   134-136
   conclusion, 149-151
   cooperation in negotiation of treaties and
   agreements, 146-148
   cooperation through international
   institutions, 148-149
   empirical reality of statecraft, 140-142
   encouragement by united nations, 139-140
   generic variants of mandate, 137-138
   guidelines and principles, 132-134
   introduction, 123-125
   mandate for cooperation in OST, 128-131
   mandate of international cooperation, 127-128
   moon treaty, 131-132
   nature of space law, 125-126
   period of superpower détente, 142-144
   space law prior to OST, 127
   superpower cooperation with other
   countries, 144-146
   UN declaration on international cooperation,
   138-139
International Space Station (ISS), 37, 41, 46,
54, 67, 72, 149
International Telecommunication Union,
(ITU), 216
Iraq, 80
Islam, 39
Israel, 81

Japan, 63, 78, 84, 85, 89, 90, 94, 104, 114, 136,
145, 149, 157, 209, 213, 228
Jupiter, 159, 175
Jus Cogens of Space Law: A Proposal, 1-26
   and state sovereignty, 9-11
   appraisal, 24-26

INDEX

concept of customary international law, 3-5
concept of, 7-8
fiduciary aspect of, 13-14
freedom of access to outer space, 16-18
international cooperation as cardinal principle, 20-24
introduction, 1-3
moral element in, 11-13
OST as customary international law, 5-7
outer space as province of mankind, 15-16
space law, 14
state responsibility to humanity, 18-20

Kasturirangan, K, 64
Kissinger, Henry, 143

Law of the Sea Treaty, 15, 171, 181, 195
Lord Krishna, 165

Madras, 65
Malaysia, 86, 213
Maldives, 104, 106, 116
Mangan, James Thomas, 167
Mars, 63, 74, 78, 157, 168, 175
Marxism, 36
Mauritius, 86, 104, 106
Mecca, 39
Medvedev, Dmitry, 92
Menter, Martin, 194
Mining of Asteroids: A Legal Analysis for Effective Governance, 155-200
appraisal, 189-190
asteroid selection, 162-163
claims to celestial bodies, 165-169
composition of asteroids, 159-161
conclusion, 200
economics of asteroid mining, 161-162
existing legal provisions, 190-196
asteroids, movable or immovable property, 191-192
common heritage of mankind, 192-196
moon agreement, 190-191
outer space treaty, 190
introduction, 155-157
legal analysis, 169
material harvesting, 164-165
mining considerations, 163-164
moon agreement, 178-188
applicability of, 180
common heritage of mankind, 181-184
freedom of access, 184-186
freedom of scientific investigation, 187-188
hints of permissibility, 186
preamble to agreement, 179-180

province of mankind, 181
nature of asteroids, 157-158
need to establish strong governing authority, 197-199
need for review of moon agreement, 196-197
outer space treaty, 169-178
freedom of access and investigation, 173-174
hints of permissibility, 174-175
non-appropriation in outer space, 172-173
not prohibited is permitted, 177-178
preamble to treaty, 170-171
province of all mankind, 171-171
public-private cooperation, 175-177
population of asteroids, 158-159
state responsibility for national activities, 188-189
Mitigation of Space Debris, 2007, 22, 134, 135, 148 218, 221, 223
Mongolia, 86, 146
Moon (Moon Agreement), 6, 15, 16, 17, 18, 19, 21, 34, 44, 45, 130, 131, 132, 137, 148, 156, 157. 169, 172, 173, 174, 175, 176, 178, 179, 180. 181, 182, 183, 184, 185, 186, 187, 188, 189. 190, 191, 193, 195, 196, 197, 200, 205, 216. 226
Moscow, 36, 85
Mother Mary, 37
Muslims, 39, 91, 193, 198
Myanmar, 86, 104, 106, 117
National Aeronautics and Space Act, 1958, 177
National Aeronautical Space Agency (NASA), 40, 75, 93, 111, 135, 144, 145, 157, 158, 164, 168, 187, 188, 211

Nehru, Pandit Jawaharlal, 63
Nemitz, Gregory, 167, 168
Nepal, 104, 105, 106, 116, 117
New Delhi, 104, 118
Nigeria, 1117
North Korea, 89, 213

Obama, Barack, 93, 141, 177
Office for Outer Space Affairs (OOSA), 55, 128, 139, 198
Organisation of Islamic Countries (OIC), 105
Ostro, Steven, 162
Outer Space Treaty (OST), 2, 7, 15, 16, 19, 20, 24, 31, 32, 33, 34, 42, 47, 48, 49, 52, 53, 56, 88, 123, 127, 128, 130, 131, 132, 135, 137, 166, 168, 169, 171, 173, 174, 178, 180, 181, 182, 184, 188, 190, 200, 215, 219, 221, 222, 223, 225, 226, 229

Pakistan, 89, 90, 91, 103, 104, 105, 106, 114, 115, 117
Pentagon, 89
Peru, 86
Poland, 146
Polar Synchronous Launch Vehicle (PSLV-C9), 73, 74, 112
Polar Synchronous Launch Vehicle (PSLV-C15), 74, 112
Prevention of Arms Race in Outer Space (PAROS), 80, 212

Radhakrishnan, K (Dr), 71, 72, 74
Radio Regulations, 135, 216, 219
Republic of Korea, 85, 86, 104
Research & Analysis Wing (R&AW), 118
Romania, 146
Russia, 11, 63, 72, 73, 75, 77, 84, 85, 89, 91, 92, 93, 104, 113, 141, 142, 146, 208, 212, 219

Sarabhai, Vikram (Dr), 63, 66, 110, 111
Saudi Arabia, 39
Sagan, Carl, 162
Sharma, Rakesh, 91
Sikhs, 38
Singapore, 82, 86, 213
South Asian Association for Regional Cooperation (SAARC), 86, 104, 108, 110, 111, 112, 113, 117, 118
South Korea, 89, 90, 94, 104, 114, 136, 213
Space Policy of India: A Few Pointers, 63-95
   conclusion, 94-95
   creditable achievements in space technology, 74-76
   evolution of space policy, 65-66
      earliest articulation of space policy, 66
      india's early interest in cosmos, 65-66
   formal iteration by government, 69-71
   futuristic vision, 71-74
   international implications, 86-94
      India and China, 87-89
      India and other asian competitors, 89-91
      India and Russia, 91-92
      India and US, 92-94
   introduction, 63-64
   new strategy mandates, 76-86
      carving a commercial niche, 80-82
      cooperation in support of self-reliance, 82-84
      emerging security imperatives, 78-80
      international cooperation, 84-86
      scientific probes in service of humanity, 77-78
      space applications for socio-economic development, 76-77
   threshold and progress, 67-69
      economic constraints, 68-69
      technological and intellectual resources, 67-68
Sri Guru Granth Sahib, 65
Sri Lanka, 104, 105, 116, 117
Sweden, 84
Syria, 86

Thailand, 86
Twitter, 101

United Kingdom (UK), 85
United Nations (UN), 1, 2, 5, 6, 9, 13, 17, 22, 23, 24, 32, 47, 50, 51, 52, 55, 56, 57, 59, 66, 86, 94, 111, 124, 127, 128, 131, 132, 133, 134, 135, 137, 138, 139, 140, 146, 147, 148, 167, 169, 173, 181, 183, 186, 194, 195, 196, 197, 198, 200, 205, 207, 209, 212, 215, 217, 218, 219, 221, 222, 223, 225, 226, 227, 228
   Charter of, 2, 5, 13, 117, 68, 127, 167, 173, 181, 217
   Institute of Disarmament Research, 209
   Security Council, 7, 217
United States (US), 23, 40, 41, 45, 46, 63, 75, 81, 84, 85, 87, 88, 89, 90, 92, 93, 95, 104, 113, 114, 117, 135, 136, 141, 142, 144, 145, 146, 149, 157, 163, 166, 167, 168, 171, 177, 178, 187, 188, 208, 209, 211, 212, 213, 220, 226, 227
USSR, 11, 84, 87, 91, 105, 143, 145, 146, 149

Venezuela, 86
Vienna Convention, 8, 19, 140

Wellington Treaty, 195
Weapons of Mass Destruction (WMDs), 213, 223
World Space Authority (WSA), 198, 199, 200
World War I, 9

Yahoo, 101